S0-BWX-680

PLANNING FOR CHANGE

A COURSE OF STUDY IN ECOLOGICAL PLANNING

PLANNING FOR CHANGE

A COURSE OF STUDY IN ECOLOGICAL PLANNING

James A. Lahde, Ph.D.

Southfield Public Schools
Southfield, Michigan

TEACHERS COLLEGE, COLUMBIA UNIVERSITY
NEW YORK & LONDON 1982

To my Mom and Dad, who gave me the greatest gifts a child could receive: the need to question, the desire to learn, and the ability to persevere.

Published by Teachers College Press, 1234 Amsterdam Avenue,
New York, N.Y. 10027

Library of Congress Cataloging in Publication Data

Lahde, James A., 1937–
Planning for change.

Bibliography: p.
Includes index.
1. Land use—Planning—Environmental aspects.
2. Urbanization—Environmental aspects. 3. Ecology.
I. Title.
HD108.6.L33 333.73'17 82-834
ISBN 0-8077-2685-0 AACR2

Manufactured in the United States of America

87 86 85 84 83 82 1 2 3 4 5 6

John Brainerd

A Brook in the City

The farmhouse lingers, though averse to square
With the new city street it has to wear
A number in. But what about the brook
That held the house as in an elbow-crook?
I ask as one who knew the brook, its strength
And impulse, having dipped a finger length
And made it leap my knuckle, having tossed
A flower to try its currents where they crossed.
The meadow grass could be cemented down
From growing under pavements of a town;
The apple trees be sent to hearthstone flame.
Is water wood to serve a brook the same?
How else dispose of an immortal force
No longer needed? Staunch it at its source
With cinder loads dumped down? The brook was thrown
Deep in a sewer dungeon under stone
In fetid darkness still to live and run—
And all for nothing it had ever done,
Except forget to go in fear perhaps.
No one should know except for ancient maps
That such a brook ran water. But I wonder
If from its being kept forever under,
The thoughts may not have risen that so keep
This new-built city from both work and sleep.

—Robert Frost

CONTENTS

ACKNOWLEDGMENTS

Much of the work for this book was done for an urban ecology class at Southfield-Lathrup High, Southfield, Michigan. I wish to acknowledge the assistance of my colleagues at Southfield-Lathrup High and Southfield Education Center who originally gave me the opportunity to get this integrated subject off the ground. Special thanks also to the many students at Southfield-Lathrup who challenged the concepts in the text and so shaped its form.

I want to express my enduring appreciation to John Brainerd, whose editing gave the book whatever stylistic form it now has, and to Charles Leman, who introduced me to planning and gave me the opportunity to work in the field.

I wish to thank especially Virgina Finney, whose patience as a listener and enthusiasm for the ideas expressed in this book fueled my efforts to continue through bad times.

For their comments on the early drafts of the manuscript I am grateful to Don Maxwell, Dorothy Cox, and Robert Long.

INTRODUCTION

Throughout history, societies have struggled to achieve a balance between individual rights and social responsibilities. In the short history of our country, individuals, families, community leaders, and businessmen have been relatively free to express their cultural differences and individual needs. Given unlimited opportunities and a rich and varied resource base, we have become a wealthy, prosperous nation.

The benefits have truly been many, but the advanced technologies and institutional machinery that have permitted such prosperity have also demonstrated that we exist on a finite planet with limited space and resources to be shared. Sharing demands cooperation, and cooperation on such a large scale demands planning. This is especially true in our urbanized areas, where concentrated populations and economic forces combine to generate severe ecological and sociological problems. These problems represent a unique challenge for the generation of young people who will have to solve them and help maintain a balance in a rapidly changing and shrinking world.

Today, many private and government agencies are beginning to look at land problems from broader points of view. Community planning is an accepted practice in the United States, with regional planning, watershed planning, and even airshed planning acceptable strategies in solving environmental problems. Students can specialize in city planning and urban and regional planning, as well as the more traditional landscape architecture, at our colleges and universities.

Though past efforts have been commendable, the overall effect has been to shift the thrust of regulating land-use changes from the local individual to the local group. In the process there has been very little gain for society as a whole. Both types of planning maintain a parochial, short-term interest. There must be incorporated into the existing system a scheme that will balance individual needs with needs of all people and future generations. What is needed is a personal planning ethic: a deep-seated recognition of our individual responsibility to the world system as a whole. Most land problems are generated at the local level. They are everybody's problems. Every home, cottage, street, parking lot, business, and school has an impact, which, when measured on a collective basis, constitutes a serious threat to society. Planning, in this respect, should not be looked upon as an individual commitment to a planning profession but as an individual commitment to a personal-social life style.

This ethic can only be brought to fruition if our educational system recognizes the importance of an ecological approach to environmental problem solving. The solution to land-use problems requires a basic knowledge of the sciences, social sciences, and the arts. Thus, to reach the majority of citizens, the job must be integrated into our secondary schools. *Planning for Change* helps meet this need. The text is designed to be used at the high school level or in the fields of college-level adult education for community leaders and developers. Content materials were organized to permit the text to be taught as a semester or as a full-school-year course of study, help bridge the gap between the social and natural sciences, and help educators organize their teaching of a multidisciplinary study.

In the first portion of the text, a brief overview of the existing planning process is presented, along with the land-use problems and forces that prevail in our society and generate these conditions. In chapter 3 a basic ecological planning process is presented as an alternative to the existing situation. Using the ecological process as a guide, chapters 4 and 5 discuss the basic graphic skills and mapping techniques necessary for identifying and illustrating data collected on a site. Chapters 6 through 10 are concerned with two areas: collecting statistical data on populations to determine whether a proposed land change will be profitable or functional, and the techniques of collecting data on the natural resources of a site. Chapter 11 describes several methods of evaluating all the data that is related to a potential change in land use and suggests several methods that can be used to develop a graphic framework that illustrates areas to be preserved and or developed. Chapter 12 discusses the final design of a master plan. How to determine the best layout of developmental features, how to evaluate the layout for problems, and how to illustrate the final layout design are the major concepts discussed in this chapter.

Throughout the text it is assumed that a class or personal land-planning project will be carried out as part of such a course of study. The *Planning for Change Activities Manual* will help coordinate activities for such a project. Brief but effective graphic communication skills are required to present data collected on a site and to illustrate the many concepts or ideas that may exist in one's mind. Both the text, with its accompanying illustrations, and the activities manual are designed to maximize this type of communication effectiveness. I have made a conscious attempt to show a variety of graphic styles that are within the readers' "artistic" capabilities.

This book, then, should serve as an introduction to a philosophy of personal planning. Virtually all the information and techniques contained in the following pages are treated in greater depth in the literature from the fields of ecology, market analysis, landscape architecture, limnology, soil science, wildlife management, forestry, civil engineering, hydrology, planning, and human geography. It is my sincere hope that the concepts and skills herein will serve to stimulate further exploration into the political arena of land-use issues.

PLANNING FOR CHANGE

A COURSE OF STUDY IN ECOLOGICAL PLANNING

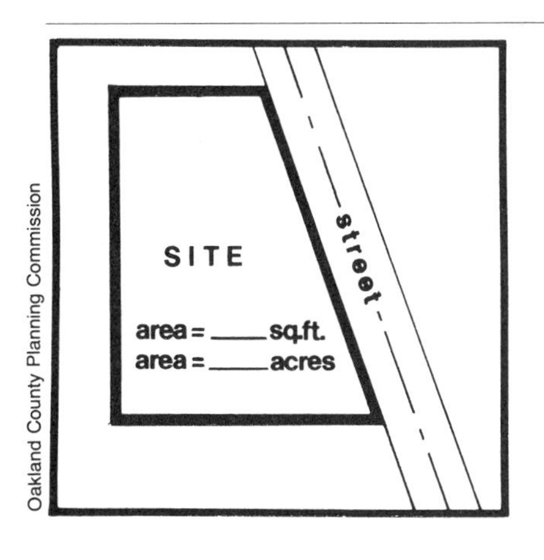

Oakland County Planning Commission

1 PLANNING PROCESSES

We have the ability to alter our environment and even create new environments. With a power that often rivals nature's, we have made mountains, changed the course of rivers, and replaced forests with factories and homes.

We have made tremendous progress in our "mastery" of nature. But as our population grows and our demand for land increases, we also increase environmental problems such as air and water pollution, traffic congestion, urban blight, indiscriminate use of pesticides, flooding, inadequate facilities for outdoor recreation, destruction of many wildlife species, and the loss of prime agricultural land.

Many of these problems are the result of unplanned or poorly planned changes to a piece of land. When homes are built, shopping centers developed, and highways constructed, certain conditions come into existence that directly or indirectly affect the environment and create problems for society.

This chapter deals primarily with existing planning processes, how they developed, how they function, and some of the problems and trends that result.

LAND PLANNING

Land planning is a systematic attempt to minimize the adverse effects land changes have on society and environments and to maximize human benefits. On a personal level, we are all planners. But over the years, a number of planning professions dedicated to giving society a more beautiful place to live have developed. Architecture and landscape architecture are examples of such professions.

Other professional planners are city and regional planners. These assist local communities or regional governments, such as county and state governments, in the design and development of their lands and waters. In this book we are primarily concerned with the planning of land within the framework of community regulations. The procedures, however, can be applied to all land changes. For example, the procedures are appropriate for the design and landscaping of back yards or school courtyards.

Oakland County Planning Commission

FIGURE 1.1. Early platting standards provided for the subdivision of land into equal-sized lots.

PLANNING LEGISLATION

State and local governments have devised a system of legal regulations and procedures to control land-use changes. Three basic types of regulations pertaining to land development are building codes, platting regulations, and zoning laws. *Building codes* set minimum standards for construction of a building; areas covered include the type of wiring, lumber, and plumbing materials that can be used in construction. *Platting regulations* guide the subdivision of land into legal separate units. Platting is the subdivision of land into proposed building sites with legal property lines that are defined by surveying the site. All the homes in a subdivision, for example, may be built on 100 × 100-foot lots (that were platted that size for residential purposes). In many cities early platting standards provided for subdividing land into 40 × 100-foot lots. (See Figure 1.1.) *Zoning laws* provide for progressive and gradual development of community land from existing land uses to proposed future uses. These zoning laws are what we shall be primarily concerned with because they permit local units of government to plan their community landscapes in a systematic fashion.

For example, Michigan's two enabling acts (the Municipal Planning Commission Act of 1931 and the Township Planning Commission Act of 1959) gave municipalities the power to write local zoning laws and establish a procedure for their enforcement.

LAND-USE DISTRICTS

Zoning laws give local government the power to divide the land within their legal boundaries into different land-use districts, each with prescribed standards for development.

Land-use districts are usually classified according to their major function, such as residential, commercial, industrial, office, and public. Following is a brief description of the most common categories. In small communities these categories may be all that is necessary; in large communities the classification may become more complex, with each category having many subdivisions.

Residential Districts

All family habitations are grouped in this category, which may include single-family homes, apartments, townhouses, duplexes, and condominiums. In the past these types were developed separately. Today residential developments can include a mixture of residential types. Often, if the development is large enough, a commercial shopping center may be included.

As a rule, city planners attempt to place residential developments away from noisy, busy, dirty areas. In some areas this may mean locating homes in the center of a development and placing apartments or commerce (businesses) near the busy streets to act as buffers. (See Figure 1.2.)

Commercial Districts

Retail and wholesale businesses, professional and business office facilities, entertainment establishments, as well as hotel and motel facilities, are usually found in this classification. Generally, commercial property is concentrated in central business districts that become known as the downtown area of the community. Commercial developments can also be found along city roads and highways. These developments may be scattered along rural roads or as long, continuous strips bordering busy highways. Some developments have been very successful in clustering a number of compatible businesses into shopping malls.

FIGURE 1.2. City planners attempt to place residential developments away from noisy, busy, dirty areas.

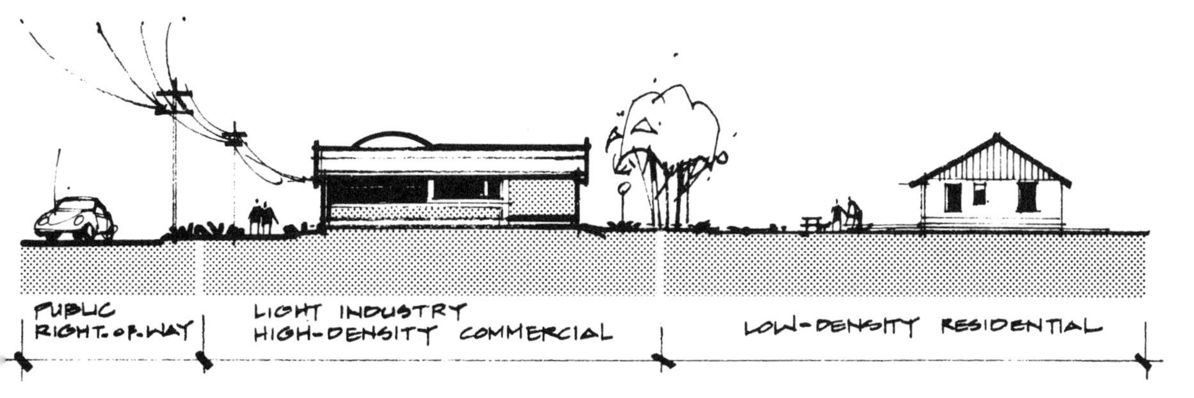

Industrial Districts

Industrial districts usually contain manufacturing enterprises, usually classified as light or heavy industrial developments. Light industry includes small, clean businesses whereas heavy industry is considered to be large, noisy, and dirty. Today heavy industry is making an effort, often successfully, to change this image.

Most community plans show that industrial plants prefer locations near major highways, expressways, waterways, or rail facilities. Light-industrial plants that are not disagreeable in any way are usually used to buffer the effects of the heavy-industrial plants from residential and commercial property.

Office Districts

Nonmanufacturing businesses are found in this category. These can include medical offices, insurance companies, real estate businesses, and so forth. These developments can be found along a community's thoroughfares or clustered in so-called professional park developments.

Public Districts (Jurisdictional Land)

Community centers, recreational facilities, cemeteries, libraries, museums, public utilities, and school sites are public districts. All transportation facilities are also included in this category, along with streams and lakes as well as roads and railroads. It should be noted, though, that only the water is considered public, not the land that abuts the water body or the land below the water body.

Agricultural Districts

This category includes areas used as crop land, orchards, and grazing land. In some communities all the land that is not designated as residential, commercial, office, or industrial is classified as agricultural. This may occur regardless of whether the land is used for agricultural purposes.

Coastal Zones

Communities bordering the sea or major lakes sometimes have strips of land with restrictions to safeguard the rural, aesthetic aspects of the shore. The State of Maine, for instance, requires such zoning by every coastal community.

Other Zones

Pennsylvania, for example, has zones preventing development of overly steep land. Flood-plain zoning is increasing in many states to prevent development of flood-prone land. In Vermont, mountaintops, with their fragile vegetation and harsh climate, have been zoned off-limits to most development.

THE PLANNING PROCESS

In this country, conscious and thoughtful design characterized early planning of such cities as Washington, D.C., Detroit, and Philadelphia. During the 1930s the federal government attempted to stimulate state planning through the establishment of the National Resources Board. By 1936 there were forty-seven state planning agencies. A few have survived. Outstanding among these few are the Tennessee Planning Commission, the Alabama State Planning Board, and the

Pennsylvania State Planning Board. There were also attempts by state legislative bodies to write enabling laws that would permit local municipalities to write their own planning laws. Michigan's Municipal Planning Commission Act of 1931 is one such enabling act. This state legislation enables communities to apply zoning legislation to their land.

Local government usually has a mayor as its administrative head and a city council as the legislative body. The city council has the power to adopt land-use laws. If adopted by the city council, the legislation allows for the formation of a planning commission.

A *planning commission* is a committee of local citizens appointed by the mayor. None of the members is required to have planning experience. By law, the duties of the commission are:

1. To make and adopt a master (comprehensive development) plan for the physical development of the community
2. To formulate, and recommend for adoption by the governing body, a zoning ordinance and zoning map based upon a comprehensive development plan
3. To advise the governing body on all matters that pertain to the physical development of the community in relation to zoning and to the subdivision of land
4. To prepare and recommend annually a coordinated and comprehensive program of public construction and improvements for the city, with priorities for their implementation
5. To consult and advise with public officials and agencies, citizens, and all public and private organizations in relation to carrying out plans prepared by the commission.

A *master plan*, usually illustrated with a map, shows what the city might look like in the future. (See Figure 1.3.) Based upon studies done by the planning commission, an area of the city is shown as a shopping area, another as an industrial area, another as a residential area, and so on.

Before the master plan is adopted, *public hearings* must be held. At these meetings, citizens are encouraged to advise the commission on the feasibility of the plan. Once a plan is adopted by the planning commission, it becomes the official land-use plan for the community. Usually the mayor and city council are not involved in the final adoption except as advisors. As a result, the planning commission exerts great power over the future of a community.

It is important to remember that a land-use plan is a plan for the future. The land-use pattern that exists at the time of the plan's adoption does not necessarily resemble the future plan. The plan may show an area as industrial, whereas it may at present be a farm. To insure the land-use changes progress in an orderly fashion from the present to the future, zoning ordinances and a review process are needed. Most municipalities hire a professional city planner to assist in these functions. Large cities have a planning department with a number of professional planners. Small communities may hire part-time planners.

Zoning ordinances set standards for each area shown on the map. For example, in a commercial area only small, neighborhood retail stores may be permitted. A store might be limited to a certain height. The number of parking spaces would be determined. Visual barriers, such as trees or a wall, may be required to obscure the parking area from neighbors. The kinds of requirements differ

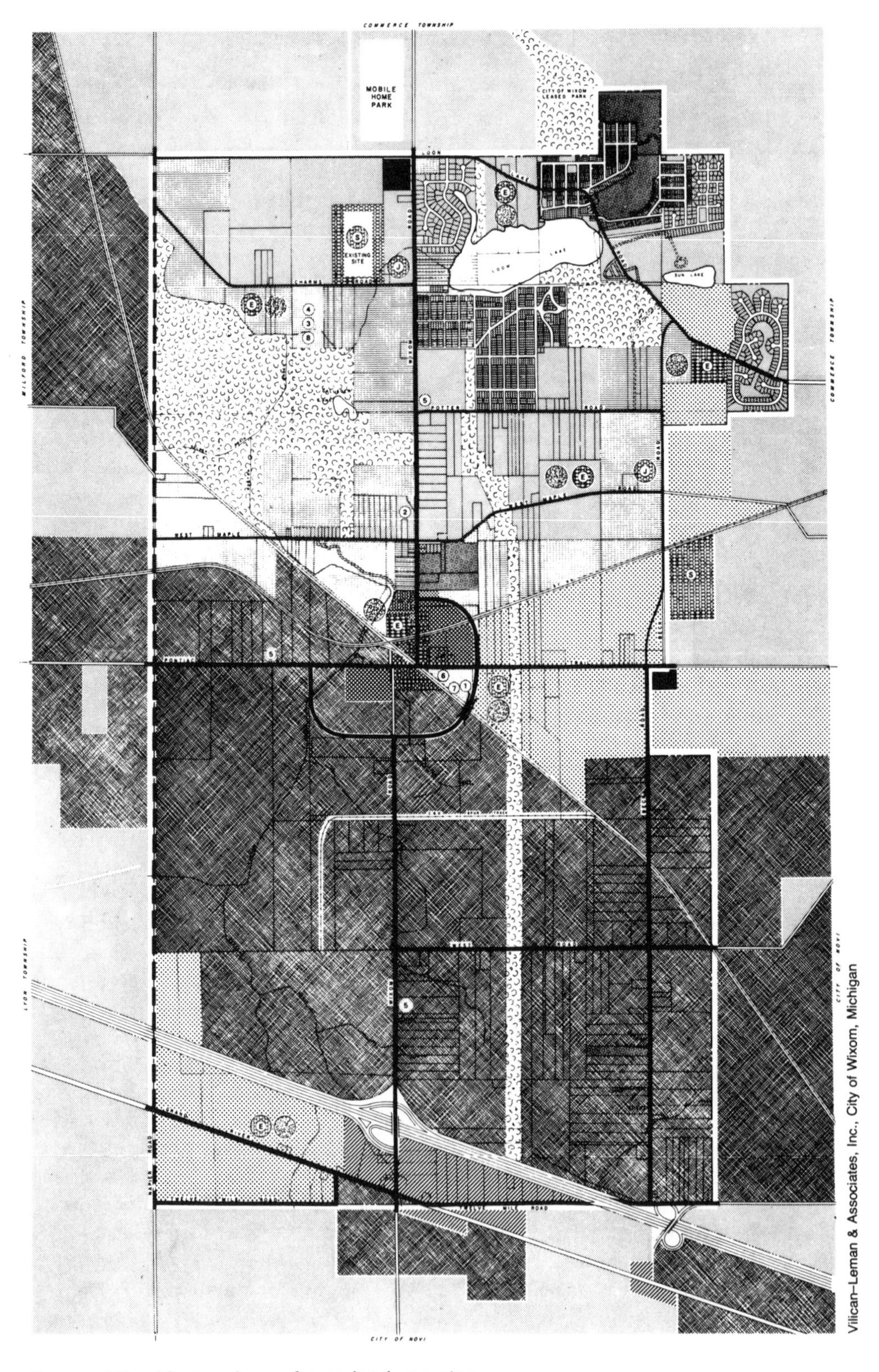

FIGURE 1.3. Master plan or future land-use plan.

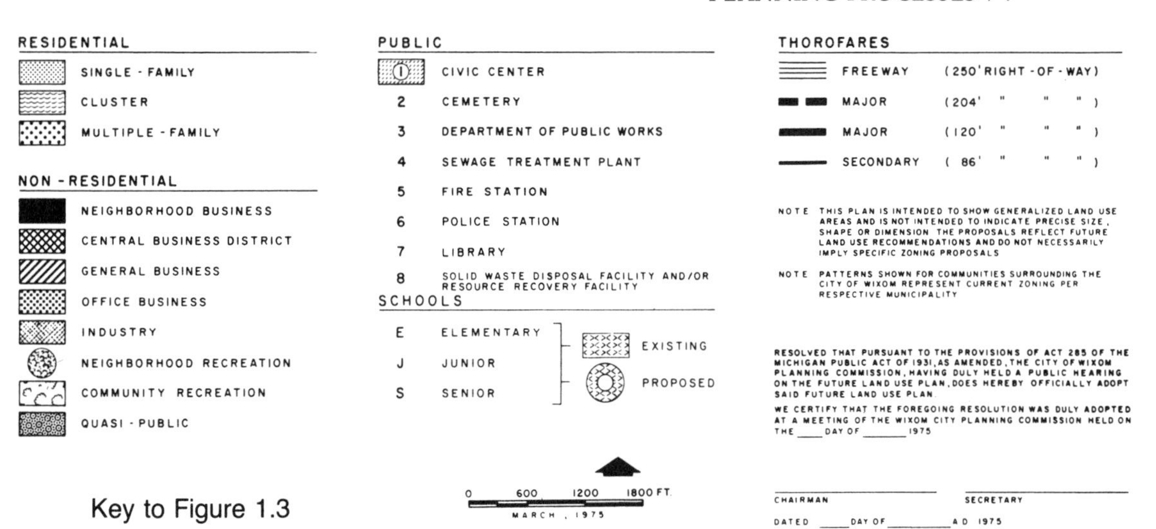

Key to Figure 1.3

from community to community, depending upon how actively concerned the community and its planners are and how they interpret the state zoning laws.

A *zoning map* accompanies the zoning ordinance. The zoning map usually shows how the parcels of land in a community are zoned at a given period. Figure 1.4 shows a zoning map of the same city that is illustrated in Figure 1.3. The zoning map is a visual aid to the zoning ordinance and is updated as zoning progresses from the present usage to the proposed master plan.

An applicant desiring a land-use change must submit plans to the community for study. This review process is fairly standard for each community. The plan is evaluated by engineers for drainage changes, building officials for construction purposes, and the planning commission for compliance with the zoning ordinance. Sometimes county and state agencies are also involved when questions arise about county roads, sewer or septic systems, soil erosion, and water flow. All these steps are not required, however, for single homes built by individuals. A local building official can usually handle most of the questions that arise in such cases.

Many times an individual proposes a land-use change that is not consistent with the city's master plan. Suppose you own a piece of land on which you plan to build a store. The city master plan may show the area as a residential district. Because your plan is not compatible with the master plan, you cannot build unless the city *rezones* the land to commercial.

To achieve that you must apply for rezoning. Your request is submitted to the planning commission for its recommendations. The planning commission will study the request with respect to its impact on the original goals of the master plan. The commission then makes a recommendation to the city council for approval or denial. At this time, public hearings are required and citizens can express their opinions on the change. The council has the final say on your request.

In a case where individuals cannot meet the restrictions established by building codes or zoning ordinances, they can apply to the *board of appeals* for relief. The board of appeals is a body of citizens appointed by the mayor that has power to modify certain local regulations on a judgment basis. For example, if

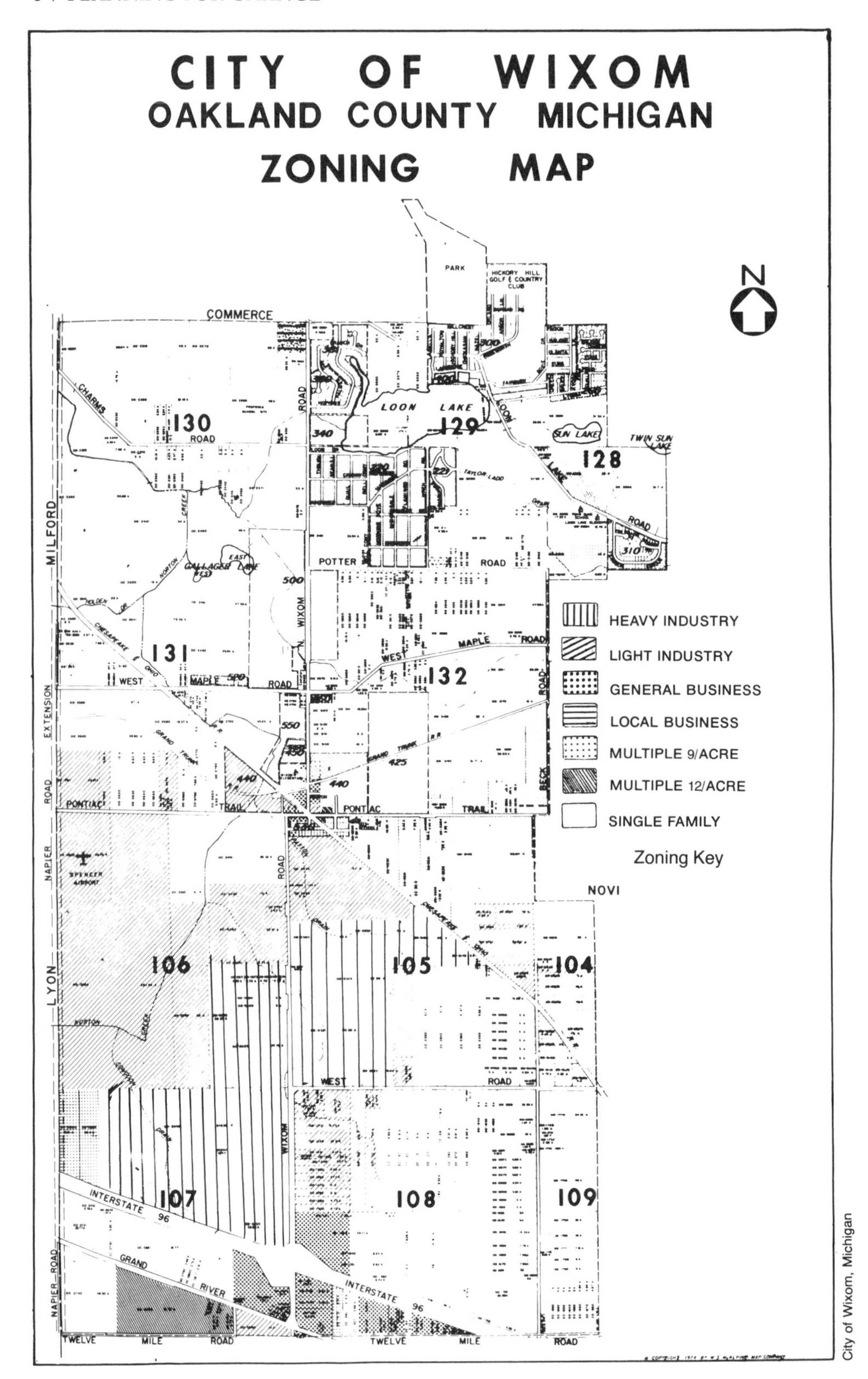

FIGURE 1.4. Zoning map.

you wish to build a home on a parcel of land that is one hundred feet wide and one hundred feet deep (ten thousand square feet), and the zoning standard for the lot's district requires a mimimum of twelve thousand square feet, you can approach the board of appeals for a *variance* (change in the regulation).

PROBLEMS AND TRENDS IN THE PLANNING PROCESS

Existing land practices have been oriented toward minimizing the *cultural* effects of uncontrolled planning practices. They have until recently largely ignored the *ecological* aspects of land changes and consequently have been destructive to the natural system in which they occur.

Lack of Ecological Understanding

Natural systems become disturbed or polluted when man's activities change the natural processes or overburden the capacity of the ecosystem to handle added materials. Traditionally, *pollution* has been defined as any departure from a state of purity. But in natural environments, it has to be thought of as a departure from natural processes. Otherwise man would have to consider volcanic eruptions and carbon dioxide production by plants as pollution.

The problems that have resulted from land-use changes are the result of man not understanding natural ecological systems and the impact this lack of understanding can have. The amount of raw sewage that bypasses a given treatment plant in New York City because of increased water runoff after a rain is testimony to the effect that man's lack of understanding can have on a resource such as the Hudson River.

Historical Causes

Part of the reason for this lack of understanding can be found in the historical nature of the planning processes themselves. Historically, land use has been a near-sacred right of the landowner. There was little room for public debate or need for systematic planning. Land use was directly related to the best use one could find for one's land.

Early land regulations reflected these values. Western water law, for example, gave an individual legal rights to water. He could dam the water for cattle or siphon it off to irrigate his crops with total disregard for his neighbor downstream. Since the 1930s there has been a national acceptance of planning using the concept of zoning community lands into land-use districts. But legal measures, like zoning, do not contribute much to the design quality of a community, especially the city centers. Zoning does little or nothing to stimulate the detailed construction of livable spaces. Only in recent years are these zoning practices being questioned and more environmentally sound and creative practices being demanded.

There is a trend today for some communities to write specific features into the zoning regulations that give a community its special character, such as greenbelts, attractive water-retention ponds, architectural designs, even the height of windowsills and the presence of vestibules in small shops. Many developers approve of this kind of zoning because it preserves those things that improve the value of their property.

Problems with Ecological Plans

As a rule, planning commission members and government officials are happy to see environmentally sound plans, but a number of social forces prevent such plans from getting to them. A major reason for the slow development of ecological planning as an accepted practice is cost. Ecological planning is like homemade apple pie: Everyone thinks it is great until the extra cost and labor become evident.

One reason for extra costs is that ecological plans usually require more open space. In urban areas land can be very expensive. With an acre of land costing thousands of dollars, a developer finds it too expensive to purchase extra land to maintain trees or protect a small portion of a stream.

Costs of planning and designing a site are a major consideration for school boards and other public agencies as well. It is difficult for government officials to justify the added expenditures involved in planning and designing a functional and attractive site. Also, the process of drafting the plans demands thorough research of a site and its surroundings. On large projects this research demands the expertise of many people. Soil scientists, botanists, hydrologists, limnologists, zoologists, planners, and demographers are some of the specialists who may be needed. This kind of study takes money. In a competitive market for land use, these expenses can be prohibitive.

More important than the direct cost is the fact that, as a people, *we often are not willing to pay the extra cost*. We have been a rich nation and could easily afford the added cost, but, unfortunately, open space, wildlife, and a constant supply of clear water have been considered frills. Most people still do not understand the long-range importance of these natural features. This fact is reflected in our land-use laws, our tax structure, and poor planning of many of our public facilities.

Platting Problems

Another problem with existing planning processes is the practice of "platting." Platting subdivides a site into proposed building sites with property lines. In the past, procedures for platting land were carried out in a rigid, linear fashion. Figure 1.5 illustrates a typical layout of homes platted in this way. Homes are constructed as close together as possible, often in a gridlike pattern up and down a street. Natural landscapes are destroyed by such practice; streams are buried or straightened out; vegetation is cut down; hills are leveled. As a result, many urban and suburban communities have taken on a sterile, monotonous atmosphere and have magnified flooding, erosion, sedimentation, and water-pollution problems with the increased runoff of water.

Fortunately, the trend today is to break away from the linear, equal distribution of land. Some developments now are laid out in a pattern to conform to the natural characteristics of the landscape, or concentrate building units on less land and thus provide for large tracts of common open space. This concentration is known as *cluster zoning*. High-rise developments, apartment complexes, condominiums, townhouses, and garden apartments can produce similar results.

Problems with the Enabling Acts

The enabling acts themselves contribute to environmental problems. The land-use acts are usually not mandatory. As a result, some communities do not

FIGURE 1.5. Traditional layout of homes often resulted in ecological problems and a monotonous landscape.

choose to adopt planning programs. Some feel they cannot afford the expense involved. The job of drafting master plans can be very technical and time consuming. It can also be costly. Because planning commission members are not aware of all the planning requirements, full-time or part-time planners are usually hired to act as consultants to the commission. Cost is also involved in writing and reproducing maps and ordinances.

While in general the enabling acts have been helpful, a major problem with them is that they are restrictive. As discussed earlier, areas in a town or city are restricted by law to specific purposes, such as homes or businesses. These areas are called districts. Early zoning regulations were often rigid documents with district lines following property lines, roads, or railroad systems. The purpose for each district often followed the principle of the "highest and best economic use for man." Little consideration was given to the people's long-term needs, which were often jeopardized because of a lack of ecological considerations. There was no protective zoning. For example, if a section of land was located near a railroad and a highway, it would be zoned industrial. The land might be a marsh with the ability to store water and serve as a sanctuary for wildlife, but such ecological factors were seldom considered. Very often such land was drained and buildings were erected on the site.

The trend today is toward more social control over land use. Planning legislation is directed toward the welfare of the total community, giving the local government more power. Many communities are adopting additional ordinances in their zoning regulations to protect important social land values. *Floodplain ordinances* are written to prevent development along rivers, a too common practice in the past. *Wetland and waterway ordinances* protect fragile plant communities whose soil is saturated with water most of the year. *Greenbelt ordinances* specify areas in the community that must be maintained as buffers between different land-use districts. In such communities, recreational areas and historical areas may also be protected in their master plan. *Coastal management zones*, such as those promoted by state law in Maine, are gaining acceptance.

Throughout the country, local communities are also attempting to limit the amount and kind of development within their borders. Expressways, sewer projects, shopping centers, and industrial complexes are often being discouraged because of the problems they bring, problems such as traffic, noise, and the expenses for more schools and more police. These attempts at limiting development have usually been successful because of the broad base of support they have among local citizens.

Along with social improvements in zoning and platting regulations, there have also been less-direct attempts to control community growth. Community rejection of proposed improvements in city sewer systems, water supplies, and roads can limit the amount of development. Limiting the size of parking lots or the number of building permits is also a common control. In some cases, citizens or developers who exhibit desirable behavior in contributing or preserving *amenities* (things that give pleasure, such as clean water, clean air, or beautiful views) can be rewarded with tax reductions. Some states have preferential property tax assessments to protect forest, farmlands, and recreational resources.

These trends to turn planning in a more socially responsible direction have led to a variety of legal conflicts. One area of conflict has resulted from the fact that changes in methods of taxing land have not kept pace with land-use changes. For example, many new laws give government the power to set standards for some private land, such as wetlands and waterways. An individual may own forty acres of land zoned industrial. Industrial land in the area sells for twenty thousand dollars an acre. Five acres of the land are swamp and marsh and the local city ordinance may, through wetland and waterway ordinances, restrict the use of this land to recreation, water recharge, or wildlife. Through the years the taxes on the land have been assessed at high industrial rates. The landowner has paid the taxes in good faith expecting to sell the land at a high profit. Now an ordinance is passed limiting his use of a portion of the land. His land becomes less valuable because of the law. Since his land has lost

Barbara Brainerd

FIGURE 1.6. Amenities: things that give us pleasure.

value, the landowner is likely to feel he has been taxed unjustly. The court battle that may result can be very costly to both parties. Often a compromise is reached, with the government gaining more social control of the land.

CONCLUSION

Chapter 1 can be summarized with a suggestion that we in America attempt to define an ethic toward the land, a standard that would guarantee future generations full benefit of the resources we enjoy. Such an ethic implies a willingness to participate at one's own expense in a commitment to a high-quality environment. It also implies an understanding of the problems and forces related to land-use planning.

Chapter 2 looks at such forces as density, taxation and the cost of city services, the influences of engineering and construction practices, and the effects of fragmented social values on land-use planning. These are powerful forces deeply imbedded in the American way of progress. It is necessary to gain an understanding of what solutions are available to alleviate the problems. We need a strategy for promoting land-use changes that are harmonious with the American method of developing land but are sympathetic to the physical and biological environment and compatible with existing planning processes.

REFERENCES

Altman, Harold, and Troost, Cornelius J. *Environmental Education: A Source Book*. New York: John Wiley, 1967.

Altshuler, A. *The City Planning Process*. Ithaca, N.Y.: Cornell University Press, 1965.

Branch, M. C. *City Planning and Aerial Information*. Cambridge, Mass.: Harvard University Press, 1971.

Conserve Oakland County's Natural Resources: Manual for Planning and Implementation. Oakland County Department of Public Works Planning Division Pub. No. 131. Pontiac, Mich.: 1980.

Friedman, John. *Retracking America*. Garden City, N.Y.: Doubleday, 1973.

Huth, Hans. *Nature and the American: Three Centuries of Changing Attitudes*. Berkeley, Calif.: University of California Press, 1957.

Lynch, Kevin. *Image of the City*. Cambridge, Mass.: M.I.T. Press, 1960.

Tandy, Clifford, ed. *Handbook of Urban Landscape*. London: Architectural Press, 1972.

U.S. Department of the Interior. *A Land Use Classification System for Use with Remote Sensor Data*, by J. R. Anderson. USGS Circular No. 671. Washington, D.C.: Government Printing Office, 1972.

Wolfanger, Louis A. *What Is Happening in Your Community?* Michigan State University Cooperative Extension Bulletin, No. F-271. East Lansing, Mich., n.d.

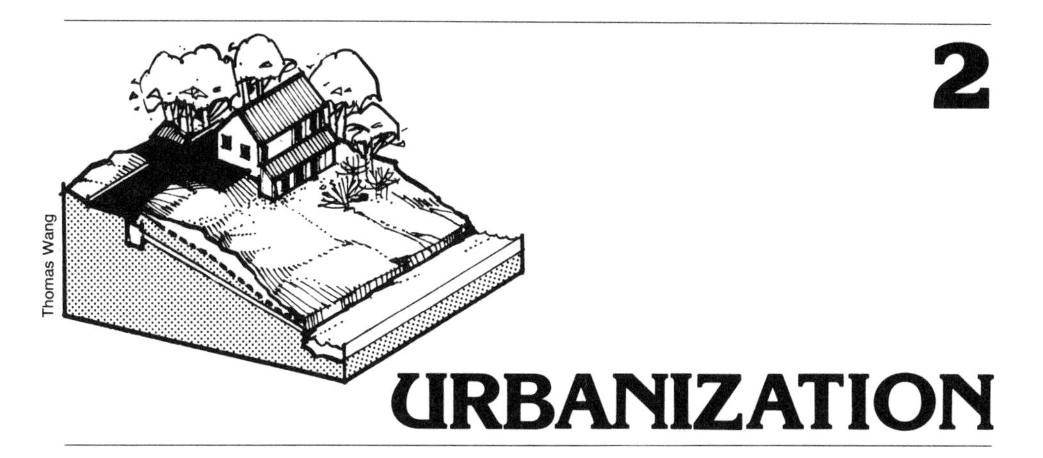

2 URBANIZATION

In chapter 1 we looked at some problems of the existing planning processes that result from inadequate land-use regulations. The whole issue of land use is much greater, though, than the lack of adequate regulation. The land-use issue is a complex social phenomenon generated by forces deeply imbedded in the social fabric of our democratic, urbanized society. Some of these forces involve recognized, standard engineering and construction practices; human rights that are considered basic democratic privileges; and overloading of limited areas with human effort and waste until the social and biological structures break down. The sources of the problems are often complex and difficult to identify. This chapter looks at some of the forces that are part of urbanization.

DENSITY AND ITS EFFECT ON LAND

The high, unplanned concentration of people on small areas of land, as in a city, is one of the major causes of social and environmental problems. Early hunting and gathering societies, made up of small bands of people, had little effect upon the environment. When game became scarce or fields began to lose their fertility, the people would simply move to another location. This gave the land an opportunity to revitalize itself.

As scientific knowledge increased, so did population. Medical science cut the death rate and improved the longevity of mankind. Technology improved the quality of life. Technological processing began to locate in strategic places where raw materials, labor, and markets were close, and whence products could be shipped to more distant markets. This technology required a large number of workers. Countries began to experience a shift in population from rural to urban areas, concentrating people on smaller parcels of land. As a result of this migration, people lost their contact with and respect for the land; cities began to grow larger with no long-range plans for guidance. This *unplanned development* and *increased density* resulted in an increase in the social and environmental problems experienced by city dwellers. Open space, associated recreational activities, and other amenities became scarcer. Urban blight, traffic congestion, and air and water pollution became realities, giving rise to the need for planning on a more systematic, long-range basis. (See Figure 2.1.)

FIGURE 2.1. Unplanned development and increased density often result in social and environmental problems.

Dave Larwa

URBANIZATION

Such a simplified view of a city suggests that density causes all the problems. Actually, many interrelated factors can combine to create environmental problems in urban and suburban areas (and more recently in rural ones too).

One of these factors is the impact of land taxes on open space, especially farmland. In most cases, rural land is owned by private individuals who pay land taxes based on present market value. *Market value* is the average price that the average citizen interested in a piece of land will pay for it. As city environments become less desirable, populations move to the fringes of the city. The process of changing rural environments to more urban ones is *urbanization*. Before the influx of people into these rural areas, the citizens pay relatively little in taxes: there is little demand for the land, and services are at a minimum. As more people move out to the fringes of the city, the competition for land increases, and therefore the market value of land increases. A farmer whose land was worth five hundred dollars an acre soon can realize five thousand dollars an acre or more by selling it at its new market.

Because a farmer's taxes are based upon the value of his property, his taxes are raised just as sharply. Soon the farmer *must* sell, for he can no longer afford to make an adequate living by farming.

Dave Larwa

FIGURE 2.2. Large increases in taxes cause farmers to sell their land because they can no longer make an adequate living by farming.

Another factor that generates problems is the cost of urbanization. Roads that were adequate for rural traffic must now be improved. The water supplies and sewer systems must be radically changed to meet the concentrated demand. Police and fire departments must increase with the population, as do crime and fire. Even recreation becomes expensive. Hockey rinks replace local ponds. The old swimming hole gives way to artificial swimming pools. Jungle gyms and treated-timber playground equipment take the place of trees and rocks. The open field is replaced with a manicured baseball diamond. All are expensive to build and maintain.

FIGURE 2.3. City services, such as recreation facilities, become more expensive in urbanized areas.

Observer & Excentric Newspapers, Livonia, Michigan

FIGURE 2.4. Apartments pay higher taxes than single-family homes.

Because the cost of urbanization is high, the burden of payment to the homeowner is relieved somewhat by taxing business at a much higher rate. Generally, large industrial plants pay the most in taxes, with offices and commercial property (stores) paying somewhat less. High-density apartments and condominiums also pay more than single-family homes. Because these multiple-family developments generate more local tax dollars, city officials encourage their growth. In the scramble for more tax dollars, most city officials have given very little consideration to long-term impacts on the environment. A spiraling effect occurs with higher density, more services, higher taxes, less open space, and so forth. The spiral continues until a saturation point is reached, with many amenities lost. Then the population moves farther out to generate the same sad condition.

Tax incentives that encourage development are also built into the federal tax structure. Many residential and commercial buildings are constructed as part of a *tax-shelter plan*. Such buildings may be built with borrowed money by investors who rent them to store chains or local businessmen. The excess money that must be paid (e.g., estate taxes, principal, and interest on loans) above the income received from the renters is a net loss to the investors. This money can be deducted as a tax shelter from ordinary income when computing taxes. As a result, a person with a high income may pay very little in personal income tax. Eventually the investor owns the land and the buildings. The investor has little to lose; besides, the development may be very successful! If this occurs, the land that the development is on and the surrounding area will increase in value. If the development does not work out for the investor, the area may not be so valuable or may become blighted. Even then, the investor does not necessarily lose.

All of the preceding factors are part of urbanization. Following is a summary of the process and the steps that result in the destruction of farmland, wildlife habitats, swamps, marshes, and other valuable open spaces.

1. The city environment becomes undesirable.
2. Portion of the city population moves out to the suburban or rural area.
3. The population shift results in competition for land.
4. The market value of land increases in the suburban areas.
5. Property tax increases result in the sale of land that in its rural condition may have important social value.

6. At the same time, city services increase and motivate governments to encourage industrial, commercial, and office development within the city to reduce the burden of taxes on private land owners.

URBAN SPRAWL

The horizontal movement of populations from the central city to the rural areas has often resulted in *urban sprawl*. In many of our larger urban areas this process has occurred to the point where suburban communities completely surround the city. Over time, the process may occur again so that several rings of suburban communities ring the central city. At first this movement began to be an important problem near the larger cities only, but today many small communities, too, are experiencing similar problems related to urban sprawl. They are beginning to realize that, even though there are many benefits to individuals from this unplanned growth, the long-term effects on society are costly.

Farmlands are the most vulnerable and very often the first social resources to be urbanized because they are relatively flat, cleared of trees, have well-drained soils, and are close to the population centers. Historically, cities developed along key locations on major river systems. These rivers served as transportation arteries and also provided necessary pockets of rich soil to feed the growing population. Expanding incomes, automobiles, and construction of roads permitted easy urban development of these rich agricultural lands. Consequently,

FIGURE 2.5. The movement of populations from the central city to the rural areas has often resulted in urban sprawl.

John Brainerd

John Brainerd

FIGURE 2.6. Estuaries are important parts of coastal regions and occur wherever fresh-water rivers join the salt water of the sea.

farmlands from Maine to Florida and from Virginia to California have been converted to shopping centers, parking lots, industrial parks, and single-family homes.

Expanding economies and rapid transportation have also permitted easy access to lands that were once considered too remote or expensive to build on. In the northeast, numerous swamps and other wetlands, regarded as wasteland, have been drained for construction and agricultural purposes. On the East Coast and Gulf Coast tidal marshes have been ditched, drained, and filled for real estate and industrial development, or for use as garbage dumps. From 1954 to 1969, New York State lost 30 percent of its wetlands, Connecticut 22 percent, and Maryland 10 percent. A similar situation exists with mangrove swamps in the more tropical coastal states.

Estuaries are important parts of coastal regions and occur wherever fresh-water rivers join the salt water of the sea. Because they permit access to both the sea and to inland waters, estuaries are especially attractive locations for commercial, recreational, industrial, and residential development. Dredging and filling generate developed waterfronts but destroy habitat for wildlife and breeding grounds for important commercial salt-water fish, not to mention destruction of much natural beauty.

In the West, especially in such places as the Palos Verdes Hills near Los Angeles, California, construction on hillsides has resulted in landslides. The long-term, progressive destruction of surface vegetation and planners' inability to consider such variables as soil composition, subsurface geology, groundwater, soil moisture, and climatic conditions have caused tremendous economic and personal hardship.

CONTRIBUTIONS OF ENGINEERING AND CONSTRUCTION PRACTICES

Flooding, drought, and the deterioration of water quality are serious problems in most regions of the United States. These problems are often generated and magnified by traditional engineering and construction practices. To understand the problem, one must understand how water works within a natural system. (See also chapters 9 and 10.)

When it rains on a wooded hillside, the force of the rain is broken by trees and other plants serving as ground cover. The leaf cover and topsoil act as a giant sponge to soak up the water. The water is slowly released back to the air, to the plants, or into a nearby stream or pond. In humid regions, some of the water moves slowly down through the soil to the water table, the top of a zone of saturated soil. (See Figure 2.7.)

Throughout the process, plant and animal life in the soil and in the water are supplied with a proper amount of water throughout the year. In the process of building farms, factories, highways, and homes, a predictable chain of events takes place that leads to water problems. Plant cover is removed from the land. Often cement, asphalt, and rooftops replace the vegetation. There is less insoak of water. As water falls on the land, an increased volume will run off the surface into nearby waterways at an increased rate of flow. This accelerated volume causes rivers to rise rapidly, lake levels to fluctuate, riverbanks to erode more quickly, and large amounts of dissolved and suspended materials to be deposited on the bottom of streams and lakes. (See Figure 2.8.)

Consider the community of Johnstown, Pennsylvania. In 1889, an earthen dam broke, sending water from the Conemaugh River rushing down the valley toward Johnstown. Two thousand people died and $17 million worth of damage resulted. In 1936, runoff water killed 23 persons and there was $41 million worth of damages. In 1977, a rainstorm resulted in 101 deaths and over $350 million worth of damages, but the city still remains.

Studies on the effect of magnified runoff have been carried out on several river basins in the Boston metropolitan region, where considerable urban growth occurred on the floodplain between 1952 and 1972. Their findings

FIGURE 2.7. Natural cover absorbs water and slows water runoff, keeping the water cool and maintaining a more constant flow and clarity of the water that enters a body of water.

FIGURE 2.8. The loss of vegetation and topsoil and the addition of roads and rooftops increase water runoff and carry more dissolved and suspended materials into a water body.

suggest that floods not only occurred with increased frequency but that flood levels that had previously been reached only once every fifty or one hundred years could now be expected within a five- or ten-year cycle.

The dissolved and suspended materials in question consist of soil components (e.g., gravel, sand, silt, clay, and humus). Streets and rooftops also contribute a considerable load of suspended materials, such as dog feces, dust, salts, oils, and litter. This material can destroy the character of the stream bottom, clog the breathing apparatus of stream organisms, and diminish the photosynthetic capabilities of plant life. Organic matter adds to the nutrient supply of the water body. Some of the nutrients are readily used by plant life, thereby increasing the population of organisms in a stream. Some of the organic materials must undergo further decomposition by bacteria. In the process, bacteria use oxygen and reduce the oxygen supply for other aquatic life. The litany of events that can occur is almost endless, with one action causing reactions all through the web of ecological processes.

Modern road construction contributes to the *eutrophication* (enrichment of a lake with nutrients) of lakes and the deterioration of streams. Traditionally, accelerated surface runoff is directed into concrete pipes buried underground. Today, many streams are restricted on all sides by concrete, stone riprap, or even old car bodies. Water flows rapidly through these structures, eventually ending up in a body of water. This increases the runoff, creating a flushing type of action, cleaning the city but multiplying the problems of flooding and sedimentation downstream.

Engineers and contractors also have often tried to straighten existing streams which tend to cause problems by meandering. Such channelization may have its benefits locally, but changing a stream's energy in one place can cause havoc downstream.

FIGURE 2.9. Restructuring a stream's channel is a common engineering practice.

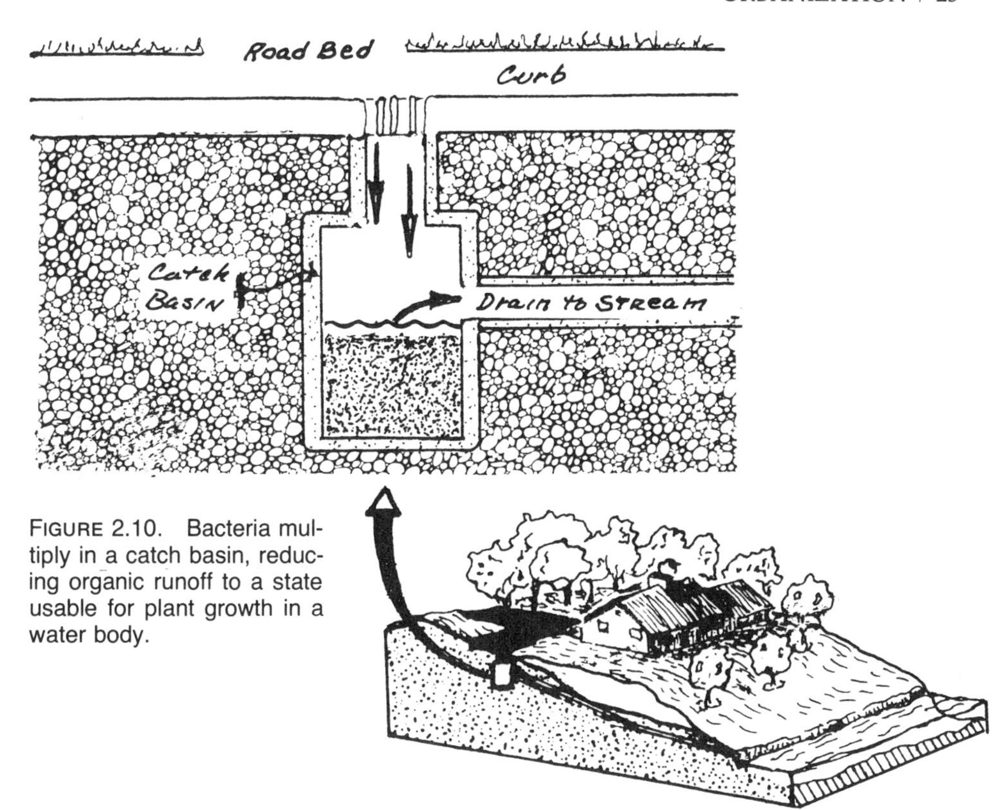

FIGURE 2.10. Bacteria multiply in a catch basin, reducing organic runoff to a state usable for plant growth in a water body.

Another practice that contributes to water problems is the use of catch basins. (See Figure 2.10.) A *catch basin* is a collecting chamber for debris washed through street grates into the storm drain. Theoretically, the catch basin will prevent the pipes from becoming clogged. The problem with a catch basin is that it functions like a septic tank with respect to organic matter that is washed into it, because the chamber is ideally suited for bacterial growth. Under such conditions, bacteria decompose the organic material into basic elements, which are then washed out when a rain occurs, adding to the nutrient supply of a water-body. Also, the necessary periodic removal of insoluble materials such as sand adds to a city's maintenance expenses.

Much of the urban and suburban population today is still serviced by *combined storm and sewage systems*. The pipes that carry runoff water unite with pipes that carry sewage from homes. This water goes to a treatment plant. Often the treatment plant is bypassed (except for chlorination to kill disease organisms) during heavy rains because of the extra volume of water that is flushed down the pipes. This also adds to the amount of organic matter entering local bodies of water. Although the practice of combined sewers is recognized as a poor method of handling water, many of our older cities are shackled to this method because of the enormous cost of change.

EFFECTS OF FRAGMENTED VALUES

Fragmented social values contribute most to the destruction of the environment. We suffer from the lack of a widely accepted set of regulations to govern use of fragile resources such as soil, water, and vegetation. In America we have traditionally had the right to use common resources as we individually please. We have not learned to think of resources in terms of the *social costs* and *social benefits* of our actions—costs and benefits to society for now and in the future. Even with the tremendous environmental problems that confront us today, we are still lax in finding acceptable, common practices for using the resources.

The Evans Drain, a tributary of the Rouge River in southeastern Michigan, is a classic example of fragmented values. The stream starts in the northeast corner of the city of Southfield, Michigan. From there it collects water and flows in a southwesterly direction until it empties into the Rouge River near the border of Southfield and Detroit.

Where the stream starts, a subdivision was built. The stream was dammed in four places to create ponds that serve the residents of the subdivision in a number of ways. The ponds hold small boats and provide fishing and skating in season. Homes built along the stream are more desirable to own because of the visual and recreational amenities.

For approximately one and a half miles, the stream passes through the northwestern corner of Lathrup Village and three school sites. Here the stream is buried underground in cement pipes and thus cannot be used for educational, recreational, or any other purposes.

As the stream emerges from under the John Lodge Expressway to Detroit, it passes through Lawrence Institute of Technology. Here the stream is left undisturbed except for clearing shrubs and cutting grass. The stream area reflects a park like character and is used as a park as well as for natural-resource education.

After it leaves Lawrence Institute, it passes through an older subdivision of Southfield. The stream here is left in its natural state. The beauty and abundance of wildlife are the main reasons expressed for this wise use of the stream, but there are other positive values too. (See Figure 2.11.)

Each of these uses has social benefits and social costs. Some uses benefit the local citizens; the same ones may harm neighbors bordering the stream elsewhere along its route. Some of our greatest resource controversies are between upstream and downstream communities who must share a common resource, a river. For example, mountain mining, dryland irrigated farming, and urban communities are engaged in political battles for the water of the Colorado River.

CONCLUSION

A lack of insight and a fragmentation of values can be seen in all levels of our social structure, public and private. Engineers, political and industrial leaders, and the public in general all contribute to the gradual erosion of the social and natural environment. Many city halls, school sites, private residences, and commercial and industrial developments are constructed with little thought for the impact they will have on society. We have seldom been taught to think in terms of social benefits and social costs rather than short-term economic costs and benefits. Consequently, most people look at their land as a commodity to do with as they please; realistically, they should view themselves as temporary tenants with responsibilities to the wildlife, to other humans, and to future generations who must use the same resources for their existence.

One of four ponds in the Cranbrook subdivision

Entrance to underground pipes that go through Lathrup Village and three school sites

Parklike character of stream through Lawrence Institute of Technology

Natural character of stream through Tamarack Village

Dave Larwa

FIGURE 2.11. A classic example of fragmented values.

These trends have caused some community officials to look for alternative methods of planning, some techniques that will alleviate the social and environmental problems but still give the developers a return on their investments. *Ecological planning* offers viable solutions to our problems. In most cases, density and volume of business can be maintained, allowing the developer an adequate return on his investment while ecological principles protect the environment and offer the public a variety of natural amenities.

REFERENCES

Blake, Peter. *God's Own Junkyard: The Planned Deterioration of America's Landscape*. New York: Holt, Rinehart and Winston, 1964.

"Cities." *Scientific American*, September 1965.

Cleveland, G., Reid, George W., and Walters, Paul R. "Storm Water Pollution from Urban Land Activity," Preprint 1033, Annual Environmental Meeting. Chicago: American Society of Civil Engineers, 1969.

Detwyler, T. R., ed. *Man's Impact on the Environment*. New York: McGraw-Hill, 1971.

Detwyler, T. R., and Marcus, J. *Urbanization and the Environment*. Belmont, Calif.: Duxbury Press, 1972.

Douglas, D. G., and Stewart, J. R., eds. *The Vanishing Landscape: A Collection of Critical Essays on Pollution and Environmental Control*. Skokie, Ill.: National Textbook, 1970.

Flawn, Peter R. *Environmental Geology: Conservation, Land Use Planning and Resource Management*. New York: Random House, 1970.

Gottman, J. *Megalopolis: The Urbanized Northeastern Seaboard of the United States*. Cambridge, Mass.: M.I.T. Press, 1961.

Pocock, Douglas. *Images of the Urban Environment*. New York: Columbia University Press, 1978.

Sharpe, G. F. S. *Landslides and Related Phenomena: A Study of Mass Movement of Soil and Rock*. New York: Pageant-Poseidon Books, 1960.

Strong, Ann. *Private Property and Public Interest: The Brandywine Experience*. Baltimore: Johns Hopkins University Press, 1975.

Tuan, Yi-Fu. *Topophilia: A Study of Environmental Perceptions, Attitudes, and Values*. Englewood Cliffs, N.J.: Prentice-Hall, 1974.

U.S. Department of Housing and Urban Development. *Handbook of Departmental Policies, Responsibilities and Procedures for Protection and Enhancement of E.Q.* Ped., Register Vol. 38, No. 137, Pt. 3. Washington, D.C.: Government Printing Office, 1974.

Ward, Colin, and Fyson, Anthony. *Streetwork: The Exploding Schools*. Boston: Routledge and Kegan Paul, 1973.

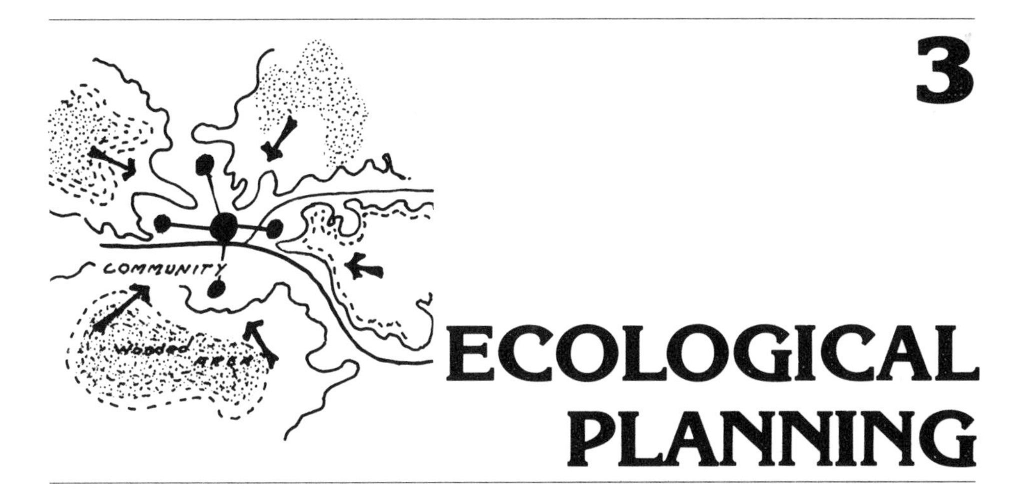

3 ECOLOGICAL PLANNING

The general pattern followed today in developing a site is to conduct an inventory of the site to determine the types of soils and the drainage pattern for water. A market study to determine the demand for the development that might go on the site may also be done. These factors are analyzed with due consideration for the kinds of units to be built. A plan is then drawn up and presented to the local planning officials. As a rule, the environmental factors analyzed (such as soil and drainage patterns of water) are studied to prevent any harmful effects these factors may have on the final construction. Very little thought is usually given to the impact of a development upon the total resource base.

To help counteract these shortcomings and those expressed in chapters 1 and 2, a more comprehensive planning approach is highly recommended; it is called ecological planning. For the purpose of this book, *ecological planning* is defined as the systematic series of studies and decisions that determine the best uses for a given piece of land, and when appropriate, that determine the most feasible arrangements of developmental structures, in order to minimize the adverse effects changes have on the natural resource base and maximize the social and natural amenities of the site. The outline that follows presents the three phases of ecological planning used in this book.

OUTLINE OF THE ECOLOGICAL PLANNING PROCESS

PHASE ONE: Inventory, analysis, and evaluation of the site, developmental needs, and the capability of community resources to satisfy these needs.

- Developmental components
    - Market studies
    - Construction features
- Community influences
    - Community services
    - Legal regulations
    - Peripheral influences
- Natural resources considerations
    - Climate
    - Topography and soils

- Vegetation and wildlife
- Water

PHASE TWO: Methodologies for planning developments, illustrating the areas on the site to be preserved and those that can be built on or developed in other ways.

- Community planning techniques
- Ecological planning techniques
- Wildlife analysis techniques

PHASE THREE: Developing the final plan, with supportive evidence of its effectiveness.

Alternative methods of development
- Laying out of developmental features
- Evaluation and testing of developmental features

Final plan with improvements
- Environmental impact statements
- Landscape and wildlife improvements
- Sedimentation and erosion-control plans
- Water improvements

PHASE ONE: INVENTORY, ANALYSIS, AND EVALUATION OF THE SITE, DEVELOPMENTAL NEEDS, AND COMMUNITY CAPABILITY

Phase One of the planning process involves determining the kind of development that is to be proposed for the site, as well as identifying, analyzing, evaluating the site and the potential of the community's resources to satisfy the developer's needs.

Developmental Components

The types of activities and construction features, such as buildings and roads, must be known if a harmonious plan is to be developed for a site. Sometimes the type of development is known, and it remains for the planner to determine the construction features and activities that will have an impact on the resource base. At other times, the owner of a parcel of land has no plan for development and would like to know what type of development would be most profitable or desirable. In this case, market studies may be required. The planner must be able to collect data through surveys or other research that will help project the future. Such data may include population growth, population shifts, buying patterns, and trends in occupations, incomes, and construction costs.

Community Influences

Community considerations relate to any limitations or potential constraints that may have an effect on, or be affected by, a change in the original site. The factors that may be involved depend upon the *size* of the development and the *program*. Most developments require some knowledge of the location of existing utilities, such as sewer, water, electricity, and telephone services, as well as the

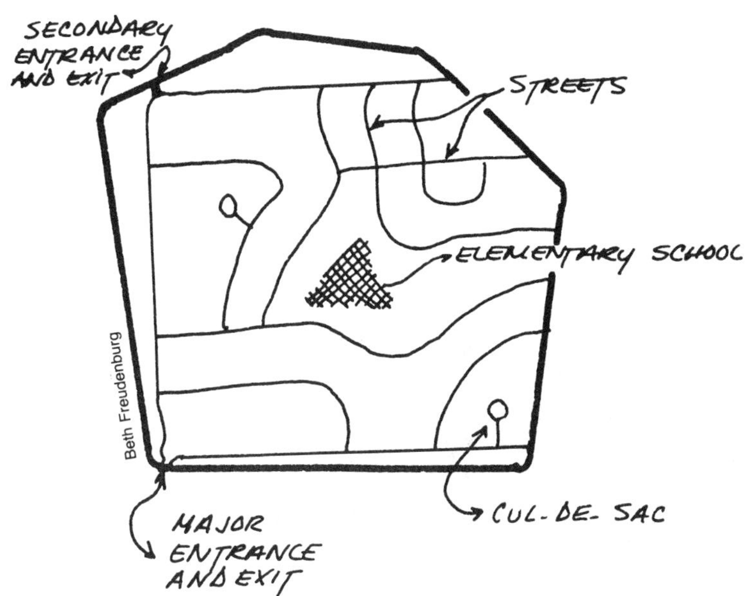

FIGURE 3.1. A simple analysis of a neighborhood.

capability of existing roads to service the development. Development also must be related to schools, fire departments, and police stations. Collectively these elements are called *community services*. (Analysis like the one in Figure 3.1—but usually more complex—are useful in understanding community influences.)

There are also important *legal regulations*, such as the existing and proposed zoning of the property and the building codes that must be adhered to.

Peripheral influences, such as character of the surroundings, can also have an impact on the development. The architectural design of nearby buildings, the character of the landscape (urban or natural, for example) and the influence of zoning by abutting communities can all have impact on the development of a project.

Such studies help planners judge the best alignment for streets to be built in a development, the availability of utilities, and the compatibility of the proposed development with abutting land uses.

Natural Resources Considerations

In this part of Phase One, as many as possible of the physical and biological factors of the site must be studied within the context of the proposed development. For example, during the inventory phase for a proposed subdivision, studies on the site should be undertaken to identify the following:

- Climatic features relating to clean air and microclimates affecting efficiency of energy and well-being of plants and animals
- Topographic features that may increase development cost, offer good vistas, or increase the potential for problems of erosion and sedimentation
- Major soil types that may affect runoff and insoak of water and health of vegetation

FIGURE 3.2. Student using transit to measure topography.

- Surface water and groundwater for water quality and seasonal fluctuations
- The vegetative communities, for instance unique stands of forest and grassland, specimen trees, and fragile and rare plant species
- The kind and relative abundance of wildlife, along with their habitats and travel routes

PHASE TWO: METHODOLOGIES FOR PLANNING DEVELOPMENTS

Too much land planning today lacks a logical, scientific method of identifying what resources should be preserved on a site and what areas are to be built upon so that a minimum of harm is done to the resource base. We have already seen how local communities plan their land by subdividing it into land-use districts. In chapter 11, several modifications of this technique will be presented using principles of ecology, focusing on those related to wildlife management. By maintaining and improving the basic resources on the site, a developer can encourage a variety of wildlife, as well as improve other benefits for humans. The numbers and health of wild animals are usually indicative of a high-quality environment for humans.

The study should suggest a pattern for development that will allow a variety of wildlife to live in and travel through the site. Figure 3.3 is a rough sketch of such a study.

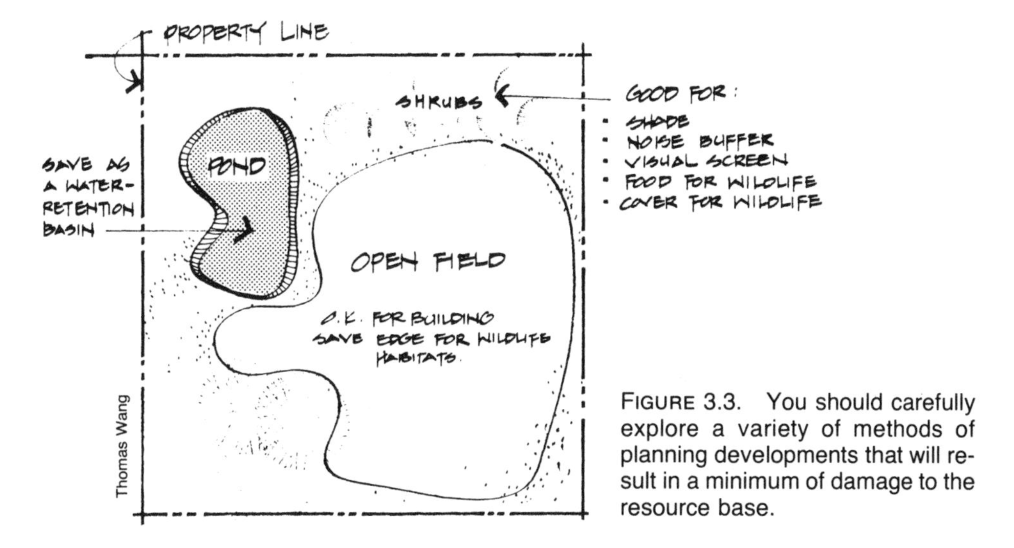

FIGURE 3.3. You should carefully explore a variety of methods of planning developments that will result in a minimum of damage to the resource base.

PHASE THREE: DEVELOPING A FINAL PLAN

In this phase of the study a final plan for the site is designed within the framework of the wildlife study and other studies identified in Phase Two. Such a plan can be evolved from alternative methods of development and a process for their evaluation, as well as from a series of studies to improve conditions for wildlife on the site.

Alternative Methods of Development

In this portion of the study, the many physical and human needs and constraints identified in Phase One and Phase Two must be evaluated to determine the best possible design. Generally this process consists of a series of brainstorming sessions where people's needs for developing a site are discussed within the framework of the site's limitations and potentials. Figure 3.4 repre-

FIGURE 3.4. Diagram of a planning process.

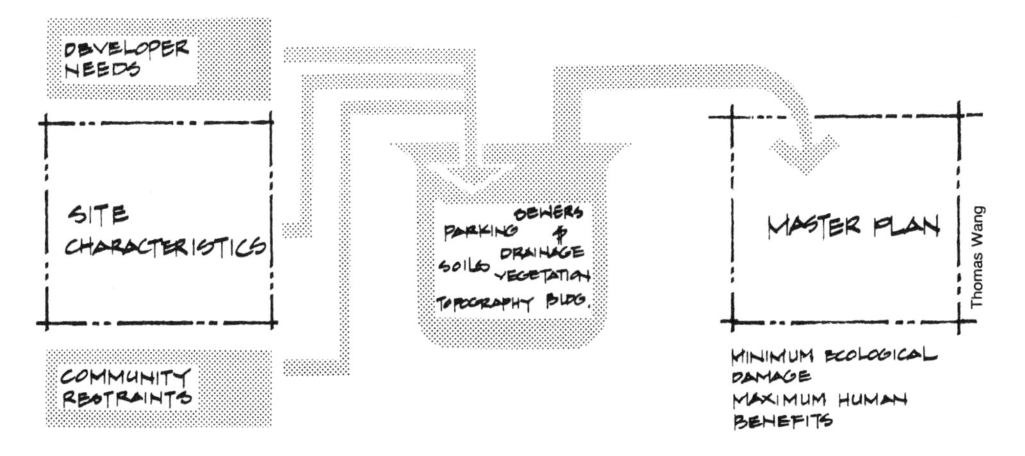

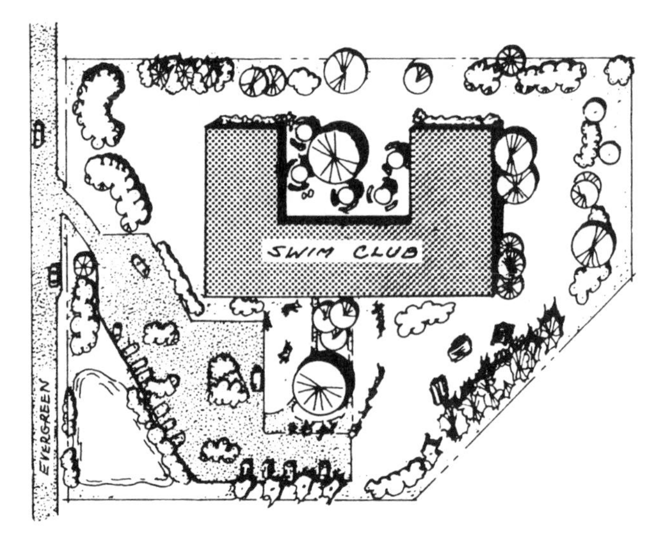

FIGURE 3.5. Master plan of a swim club.

sents a diagram of the process. All the social, human, and physical factors must be considered, and a conceptual plan must be drafted that attempts to meet the environmental and developer's needs. Many alternative layouts may be designed and tested before a final plan is agreed upon. The final plan usually shows, as realistically as possible, the layout of the development. The plan should also attempt to illustrate all major improvements proposed for the site.

Final Plan with Improvements (Master Plan)

Once all the constraints and needs are met, a final plan, often called a *master plan*, is developed. Usually it is represented by a drawing or painting. (See Figure 3.5.) Often this illustration is too general in nature to be functional as a working plan. Consequently, a number of additional, more detailed studies are included as support.

ENVIRONMENTAL IMPACT STATEMENTS

Environmental impact studies consist of reports related to the natural environment and the man-made environment. Studies of the natural environment are required on all federal projects. Such projects may include new highways, harbor dredging or filling, nuclear power plants, channeling of rivers, and many more. The purpose of these studies is to insure that environmental amenities and values are given systematic consideration equal to economic and technical factors. An environmental impact study is designed to assess in detail the potential impact of a proposed project. Some states and communities have passed laws requiring similar studies. Copies of such studies are available from almost all state and county agencies and some local ones.

Studies that are conducted to determine the impact on man-made environments are the more common type of impact studies. All developments increase the density of people in an area. This density may be permanent or long-term (housing developments), or the developments may produce temporary populations (business and schools). Studies may be carried out to determine a development's impact on such matters as

Roads and traffic
Fire and police protection
Recreational facilities
Utilities (water, sewers, gas, electricity, and telephones)
School facilities
City tax structure

Other questions include whether the increase in taxes generated by the development will compensate for the increased need for services or other problems that may result. These studies are the the specialty of urban land planners. They require an in-depth knowledge of existing and projected statistics related to local populations, taxes, city programs, and school programs.

LANDSCAPE AND WILDLIFE IMPROVEMENTS

Landscape plans describe proposed soil movements (grading) and planting designs along with topography and vegetation to be left undisturbed. Because soil is almost always disturbed during construction, most final plans have landscape improvements of some sort. Areas to be seeded with grass to prevent erosion are usually delineated. Very often the location, size, and types of trees and shrubs to be planted are included on the drawings. Wildlife improvements are too seldom treated. An ecological approach to planning such improvements *should* be included, because wildlife is a very important indicator of environmental amenities. A well-designed planting plan using ecological principles will naturally be good for wildlife.

SEDIMENTATION AND EROSION-CONTROL PLANS

Soil erosion caused by wind or water and its deposition on other land or into a body of water (sedimentation) are still some of our most severe environmental

FIGURE 3.6. Soil erosion by wind and water is still one of our most severe environmental problems.

John Brainerd

John Brainerd

FIGURE 3.7. There are many natural and structural methods of controlling erosion. Here stones and wire mesh are used to protect the stream bank.

FIGURE 3.8. A small, attractive water-retention pond.

problems. Today many states require development projects to prepare studies to show how developers will control these destructive forces both during and after construction.

WATER IMPROVEMENTS (PLANS FOR CONTROLLING RUNOFF AND INSOAK)

Some of our most severe and complex environmental problems are water-related. Sedimentation-control and erosion-control plans attempt to deal with these problems. Water must drain properly from all developments. Many states do not have effective laws in this area; some laws are poor laws or are not enforced. Today, landscape architects or engineers draw up plans that show how water will be drained from the land. A usual practice is to slope the land so that water runoff ends down in a nearby stream, drainage ditch, or storm sewer. More often, water is captured in concrete or plastic pipes that lead to nearby natural drainage systems. These methods of handling water have led to problems of increased flooding and deteriorated water quality downstream. Some communities are beginning to require water-management plans that help alleviate these problems. These plans may include streambank control measures and the development of holding ponds and sedimentation basins. In most cases the resulting ponds and small lakes are attractive and function as recreational facilities, water supplies for fire control, and wildlife habitats.

CONCLUSION

Many planning strategies are used by professional planners today. Few have a totally ecological or environmental orientation. The process outlined in this chapter and discussed in detail throughout the rest of the text, can be effective in analyzing plans proposed by developers or governmental agencies, or it can be an effective method of developing a plan for a small or a large site. To be effective, though, the process should be carried out systematically and attractively. Chapter 4 briefly discusses a number of mapping skills that a planner might use to record and illustrate site studies basic to a proposal. Chapter 5 suggests a format for developing a proposal for site plan and discusses some of the graphic techniques that can be used in illustrating such a proposal.

REFERENCES

Bosselman, F., and Callies, D. *The Quiet Revolution in Land Use Control*. Prepared for the Council on Environmental Quality. Washington, D.C.: Government Printing Office, 1971.

Cullen, Gordon. *Concise Townscape*. New York: Van Nostrand Reinhold, 1961.

Dasmann, R. F., Milton, J. P., and Freeman, P. H. *Ecological Principles for Economic Development*. New York: John Wiley, 1973.

George, Robert W., Mokma, Arnold, and Hetherington, Martin. *Ecosystem Analysis*. Michigan State University Cooperative Extension Bulletin. East Lansing, Mich.: 1973.

Lovejoy, Derek, ed. *Land Use and Landscape Planning*. Aylesbury, England: Intertext, 1973.

McHarg, Ian L. *Design with Nature*. Philadelphia: The Falcon Press, 1969.

Meshenberg, Michael J. *Environmental Planning*. Planning Advisory Service Bulletin 263. Washington, D.C.: American Society of Planning Officials, 1970.

Odum, E. *Fundamentals of Ecology*. Philadelphia: Saunders, 1971.

Perin, Constance. *With Man in Mind: An Interdisciplinary Prospectus for Environmental Design*. Cambridge, Mass.: M.I.T. Press, 1970.

Rasmussen, Steen E. *Experiencing Architecture*. Cambridge, Mass.: M.I.T. Press, 1962.

Rubenstein, Harvey. *A Guide to Site and Environmental Planning*. New York: John Wiley, 1969.

Watt, K. E. *Ecology and Resource Management*. New York: McGraw-Hill, 1968.

4 COLLECTING SITE DATA: MAPS

This chapter and those that follow have been designed to help you by describing techniques you will practice in any project you may undertake, at your school, playground, camp, or any other site in your community, now or in later life.

Much of a planner's work involves collecting data from sites to be developed. Among the ecological planner's most useful tools are various types of maps. Consequently, a planner should become familiar with the tools and skills involved in reading and understanding maps. When working outdoors it is essential to know exactly where your work is being done. Maps will also help you locate particular features such as land forms that may be of interest. Just as important, maps can serve as records or data sheets. By using maps to record information collected, the feature's location can be recalled quickly. The size and configuration of a land form and many relationships between one environmental component and another can be illustrated.

Having a variety of maps available is important in locating a city's boundaries or the major highways that help locate your site in relation to the community or region as a whole. Having a variety of maps at your disposal may also help you draw a base map of your site. One of the maps may be the exact size you need and can be copied, saving the time and money required to reduce a larger one or enlarge a smaller one.

Gather as many maps of your site as you can. At a city hall or town hall, tax assessors and departments of streets and engineering as well as planning will have maps you can consult. Surveyors and architects listed in the telephone book may have maps that include your area. Some people will make a copy to give you or sell you. Some will let you trace a map or portion of a map in their office. It is best not to borrow maps to take away; they are too easily damaged and may be irreplaceable.

PARTS OF A MAP

The purpose of a map is to represent features of the land accurately on a piece of paper. Therefore, on any map the lines, symbols, and (sometimes) colors that

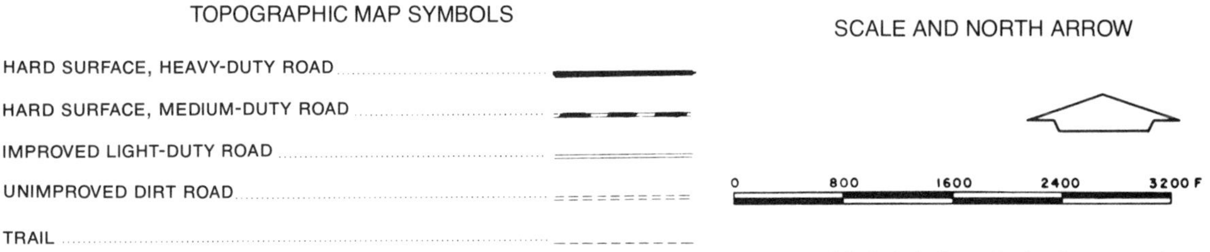

TOPOGRAPHIC MAP SYMBOLS

HARD SURFACE, HEAVY-DUTY ROAD

HARD SURFACE, MEDIUM-DUTY ROAD

IMPROVED LIGHT-DUTY ROAD

UNIMPROVED DIRT ROAD

TRAIL

SCALE AND NORTH ARROW

0 800 1600 2400 3200 F

U.S. Geologic Survey Service, Department of Interi

FIGURE 4.1. Map symbols.

represent these features are of major importance. Maps should almost always include a key (legend) to the map symbols, a north arrow, a scale, and the date; the key is an area of the map where a sample of each symbol is identified. On a road map, in addition to roads, lines may represent boundaries of political units such as states, towns, railroads, and rivers. Cities may be represented by circles or rectangles of varying size. Colors may distinguish one area from another. The north arrow (northpoint) orients the reader of the map to the proper direction relative to north on the ground. Along with the north arrow, every map should have a scale. It shows the relative distance on the map to the distance on the ground. For example, if a map shows that the distance between Nine Mile Road and Ten Mile Road is one inch, the scale of the map should be stated as "one inch equals one mile." It is best, though, to express the scale with a drawn line rather than with words, because if the map is enlarged or reduced the line will automatically stay true but the words will not. (See Figure 4.1.)

AERIAL PHOTOGRAPHS

Most maps today are prepared from *aerial photographs*, which are especially useful to ecological planners. Aerial photographs are taken with special cameras from airplanes and satellites. If you are studying a large site, an aerial photograph will give you outlines of the vegetative communities as well as up-to-date descriptions of many existing man-made features. Such information is difficult to draw accurately during surveys on the ground. Aerial photographs are very expensive to have made, but prints of most areas already photographed can be obtained from county planning offices for a small charge.

Many local planning agencies have current photographs and are willing to let you use them in their offices for a short period of time. Take with you tracing paper or tracing plastic (available at art and stationery stores).

TOPOGRAPHIC MAPS

Topographic maps have many of the features of a road map but also show landscape relief, that is, hills, valleys, and slopes. These features make topographic maps important tools for planners. Topographic maps come in different sizes relative to the area they portray. A relatively large-scale map may have 1 inch on the map equal 24,000 inches on the ground. A medium-scale map may have 1 inch on the map equal 62,000 inches on the ground. A small-scale map may have 1 inch on the map equal 250,000 inches on the ground. Beware: the smaller-scale map on the same size paper shows a larger area than does a larger-scale map. (See Figure 4.3.)

FIGURE 4.2. Aerial photograph.

Oakland County Planning Commission

FIGURE 4.3. Comparison of a large-scale map (left) and a smaller-scale map (right).

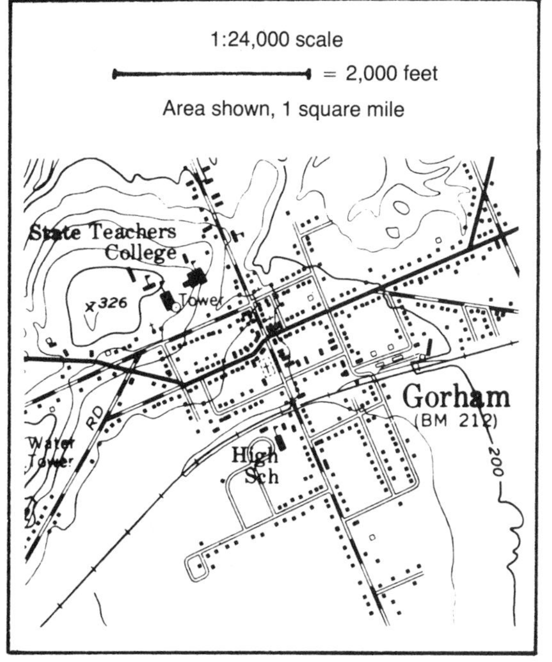

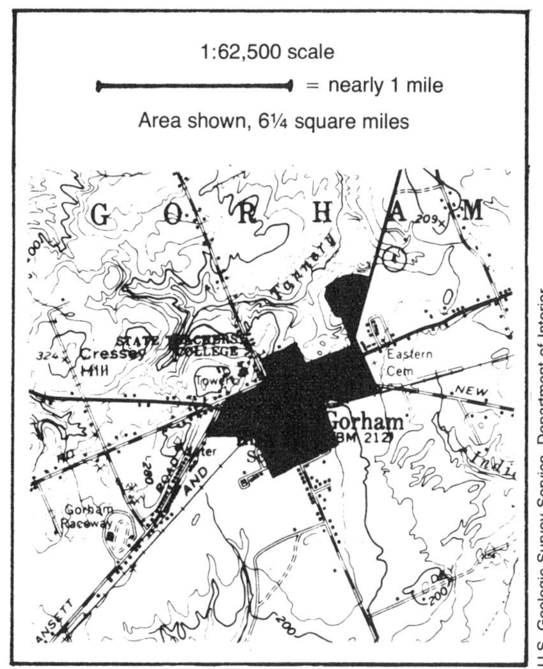

U.S. Geologic Survey Service, Department of Interior

Contour lines of topographic maps illustrate the relief of the landscape. A contour line represents an imaginary line drawn through many points on a map that have the same elevation. Elevation numbers are usually stated in feet above sea level, with sea level being zero. Practice is often necessary before a contour map can be easily read. As a planner, you should be able to perceive at a glance the landscape depicted by the contours.

Figure 4.4 shows a *perspective view* and a *plan view* of the same landscape, with contour lines in the latter to indicate elevation. Note that a river runs through the valley and empties into a bay which is protected by a sand bar. To the west of the river is a steep hill that flattens out on the top. To the east is a hill that rises gradually to a rounded peak. Notice that the contour lines representing slope differ with each hill. On the west slope the contour lines are close together. The gradual slope is illustrated by lines spaced farther apart. The vertical distance between each contour line is constant. This vertical distance between two successive contour lines is known as the map's *contour interval*—twenty-foot intervals in the case of the plan view map illustrated in Figure 4.4. Small oval contours may indicate high points. Rivers run at right angles to contour lines, as does water that runs off the surface of the land.

FIGURE 4.4. A perspective view (top) and a plan view (bottom) of the same landscape.

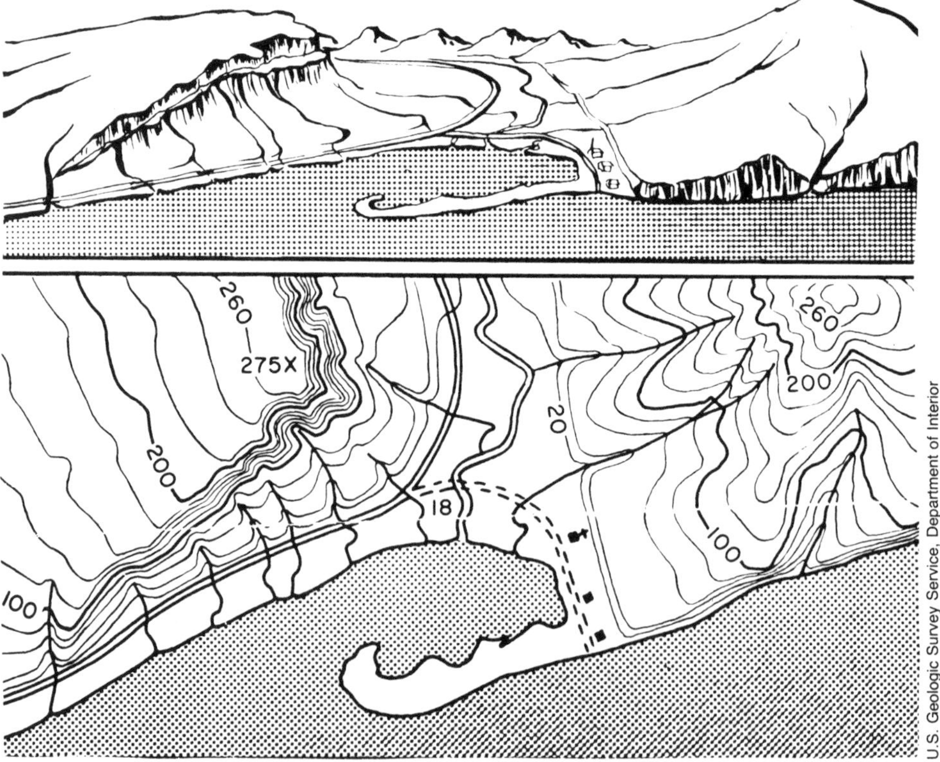

U.S. Geologic Survey Service, Department of Interior

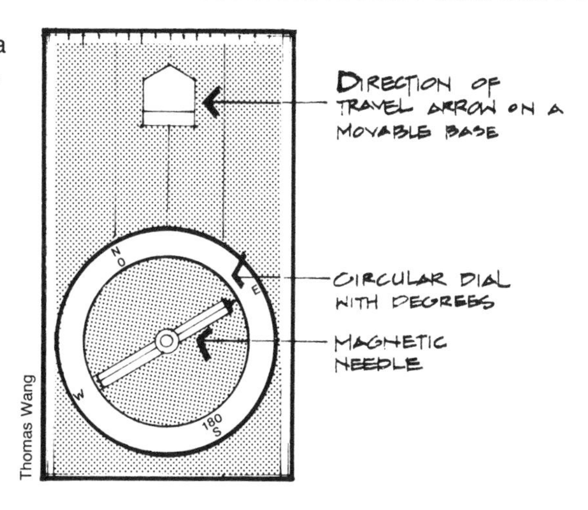

FIGURE 4.5. Top view of a common type of compass.

MAPPING SKILLS

Very often it is necessary for a planner to collect data from the site, which can then be used to construct a new map or verify the accuracy of an existing map. Reducing land features to an image on a small piece of paper requires familiarity with the compass and some methods of defining the distances between features.

The Compass

In constructing a map of a site, you must know the location of important objects on the site, such as trees, the property's boundary lines, and buildings. The most common instrument for determining the location of objects on a site is the compass. (See Figure 4.5.) Sighting with a compass enables you to determine the directions between objects so you can map their location.

The floating compass needle is magnetized so that it always points to the magnetic north. This gives you a true, constant reference point. The face of the compass is graduated into quadrants—north, south, east, and west—and into degrees. A compass used for mapping should also have a movable base with a directional arrow on it and a movable, circular-degree dial. The directional arrow on the base is used as a pointer, and the movable base represents the 360 degrees in a circle.

Shooting a Bearing

Boundary lines and land features can be located and illustrated by determining their location relative to the north point. For example, set a compass on a table in a classroom and allow the needle to settle to north. Have a friend stand several feet away, just to the right of north. You can determine your friend's exact location relative to north by pointing the direction arrow on the base of the compass toward him. (See Figure 4.6.) Next, move the circular degree dial so that zero degrees is in line with the north arrow. The direction arrow will now be lined up with the degrees to which your friend is standing relative to north. This procedure of locating an object by its degree or bearing from north is called *sighting* or *shooting a bearing* or *shooting a survey line*.

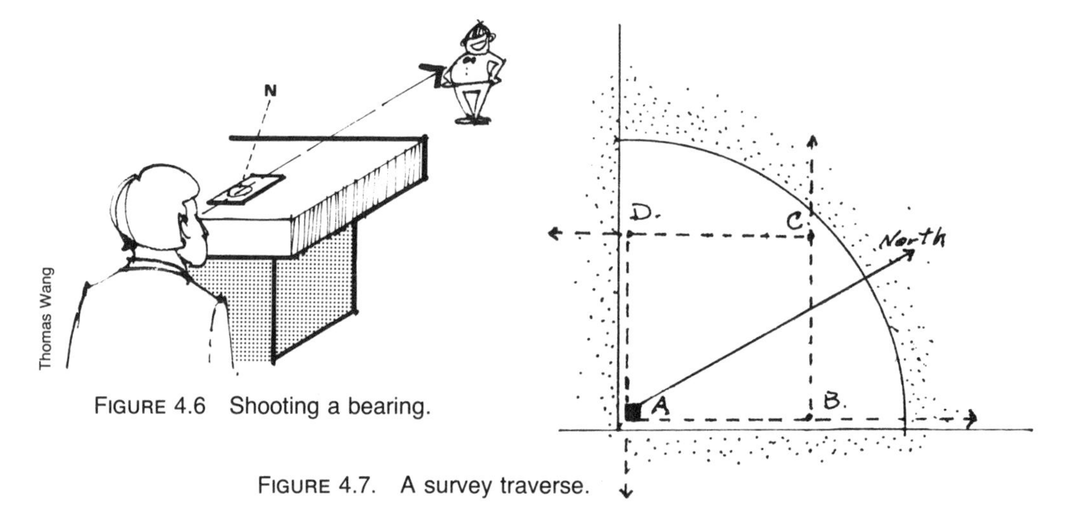

FIGURE 4.6 Shooting a bearing.

FIGURE 4.7. A survey traverse.

Distance by Pacing

In constructing a map, you must also know the distance between objects in the field. A compass bearing will supply the investigator only with the direction of an object relative to north and to the position sighted from. It does not determine the distance and scale that must be used to convert that distance to inches on a map. Distance can be measured by pacing, or by a steel measuring tape or line marked off in meters or feet (or, formerly, by a chain, a 66-foot chain-linked measuring device). Pacing, though not so accurate, is usually sufficient for most school projects. Measure by counting the number of steps it takes to reach an object, and multiply this number by the average length of each step. The average length of a step should be determined by pacing over a known distance using a normal walking pace. A school's football field is a good place to practice.

The Survey Traverse

A series of survey lines going from Point A to Point B to Point C, etc., is known as a *survey traverse*. All of the important features on a site, such as soil deposits, water bodies, vegetation, roads, boundary lines, and buildings can be mapped using the survey traverse technique. Before embarking on a survey, a rough site map should be constructed to record the site data. At the site, bearings and distances to important features are determined from a beginning reference point and recorded on the rough map.

Figure 4.7 illustrates a survey taken on a baseball field. A baseball field is a good place to practice, since the baselines are known distances that you can use to check the accuracy of your study. To map the infield of a ball field accurately, you need only record the bearings and distances between home plate and first base, first base and second base, second base and third base, and third base and home plate. Back in the classroom these calculations can be plotted (drawn accurately to scale) to produce a map of the infield.

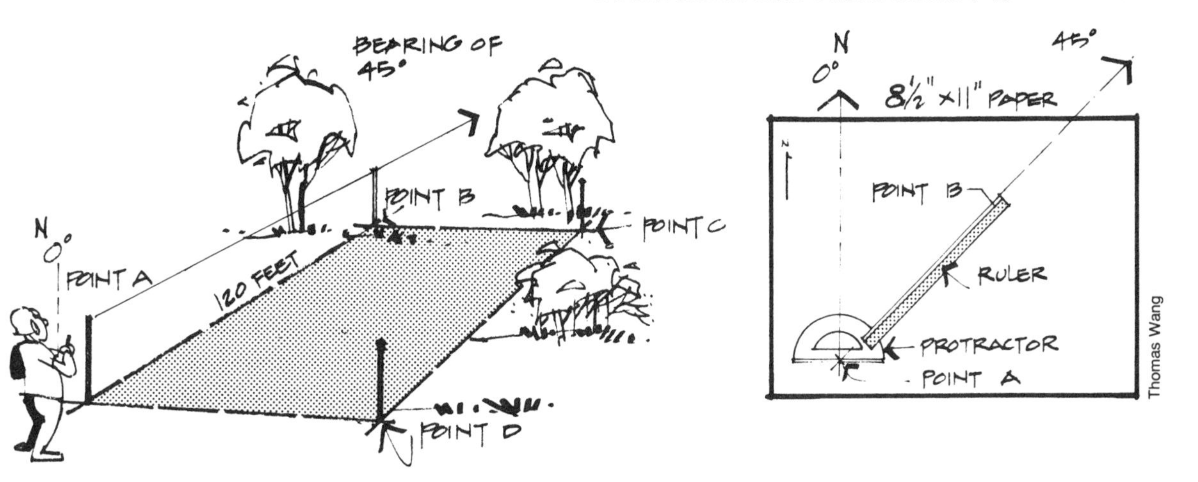

FIGURE 4.8. Plotting site data: in the field (left); in the classroom (right).

Plotting the Survey

Once the field work has been carried out, the data must be reduced to a workable size on paper, and all the data represented accurately. This operation is called *plotting the survey*. For a map to represent the features of a landscape accurately, you must decide on a scale that will reduce the site elements to a size that will conveniently fit on the paper being used. One inch on the paper to equal twenty-five or fifty feet on the ground is usually sufficient for most school projects. Once a scale is decided upon, the steps carried out in the field must be repeated, using a protractor in place of the compass and a ruler to measure and duplicate the distances.(See Figure 4.8.)

Triangulation

Sometimes it is difficult to identify the exact distance of an object along a survey line. This may occur if a lake or swamp impedes walking. Sometimes, a very general outline of an irregularly shaped stand of trees is desired. In these cases triangulation can be helpful. Triangulation also has the advantage over a survey traverse because it reduces the amount of pacing needed to establish the distances between objects. With triangulation only two points on a survey line need be accurately measured. These points should be distant enough from each other for the bearings sited to form as nearly as possible an equilateral triangle; a narrow-based, long triangle tends to give an inaccurate location of the point being sighted. From these two points, bearings can be taken to desired objects and their location identified when the bearings are plotted on paper. (See Figure 4.9.)

CONCLUSION

Having a variety of maps available, knowing how to read the maps, and having the skills to make your own maps are all important skills for the planner. They

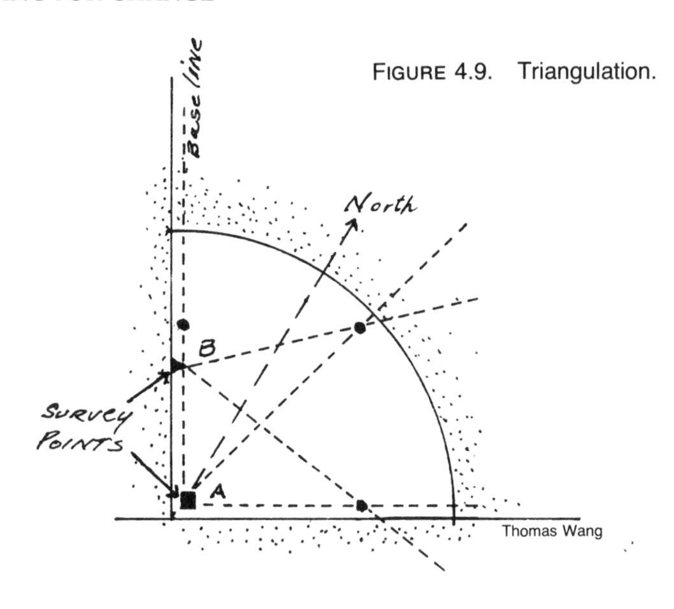

FIGURE 4.9. Triangulation.

will permit you to represent site studies accurately and present your ideas clearly and simply. Sometimes how you present an idea can be just as important as the idea itself. This is especially true if you are attempting to stimulate interest in your ideas from a client or audience.

Chapters 5, 6, 7, 8, and 10 are primarily concerned with collecting information relative to a project you may wish to develop. Attempts have been made throughout the chapters to show additional methods of illustrating data.

REFERENCES

Andrews, William A., ed. *A Guide To Terrestrial Ecology*. Englewood Cliffs, N.J.: Prentice-Hall, 1972.

Avery, Eugene T. *Interpretation of Aerial Photographs*. Minneapolis: Burgess, 1968.

Benton, Allen H., and Werner, William E., Jr. *Manual of Field Biology and Ecology*. Minneapolis: Burgess, 1972.

Berry, Brian, J. L., and Horton, Frank E. *Urban Environmental Management: Planning for Pollution Control*. Englewood Cliffs, N.J.: Prentice-Hall, 1974.

March, William M. *Environmental Analysis: For Land Use and Site Planning*. New York: McGraw-Hill, 1978.

Rogers, Everett, and Shoemaker, Floyd. *Communication of Innovations*. Glencoe, Ill.: The Free Press, 1971.

U.S. Department of the Interior, Geologic Survey Service. *Topographic Maps.* Washington, D.C.: U.S. Government Printing Office, n.d.

Way, Douglas S. *Terrain Analysis: A Guide to Site Selection Using Aerial Photographic Interpretation*. Stroudsburg, Pa.: Dowden, Hutchinson and Ross, 1973.

5

CONSTRUCTING AN ECOLOGICAL PLANNING PROJECT

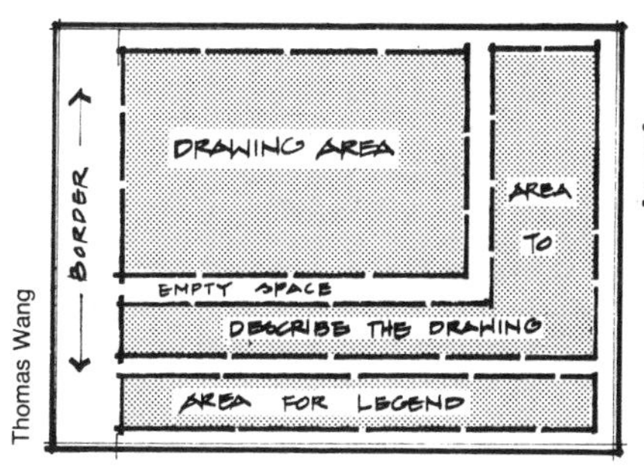

Thomas Wang

In school we are concerned with communicating with others mostly by speaking or writing; we are seldom required to communicate by pictures. In the field of planning, however, the old cliché that a picture is worth a thousand words is taken seriously.

Many of the early planners had backgrounds in landscape architecture and learned to express themselves in graphic form. In suggesting proposed changes to landscapes, therefore, they often used pictures to illustrate their ideas. Since the proposed changes were only ideas, the pictures gave the client a clear understanding of what the planners were thinking about.

Planners use a variety of graphic techniques to communicate their ideas. Chapter 4 described how site data can be collected and plotted to scale. These procedures are only first steps in illustrating a project. This chapter suggests a format for preparing land-change proposals and introduces a number of graphic techniques that can be used to illustrate the ideas involved in each phase of the study.

OUTLINE AND GRAPHICS FOR A PROPOSAL

Because of the problems inherent in an ecologically planned project, anyone attempting to improve an existing plan or "sell" a new plan should construct the plan with care, both systematically and graphically, so that all involved can clearly see the changes and the rationale for them. Table 5.1 provides an outline of a format and the graphics that can be used in constructing such a plan.

Introduction to the Proposal

The title page should have the title set prominently near the top. It should be brief, easy to refer to in writing and in talking. A subtitle may add a little more information. For example,

Table 5.1. ECOLOGICAL PLANNING PROJECT

OUTLINE	SUGGESTED GRAPHICS
Introduction	
Name of project	Title page
Booklet content	Table of contents
Project overview	Script
Location of project	Schematic map
Phase One: Inventory and Analysis	
Developmental Components	
Market Studies	Graphs and charts
Construction features	Graphs/charts/script
Community Influences	
Community services	Plan view maps
Legal regulations	Plan view maps
Peripheral influences	Plan view maps
Natural Resources Inventory	
Climate	Script
Topography and soils	Plan view and sectional views
Vegetation and wildlife	Plan view and sectional views
Water	Plan view and sectional views
Phase Two: Methodologies for Planning Developments	Bubble diagrams
Phase Three: Developing the Final Plan	
Alternative Methods of Development	
Layout of developmental features	Bubble diagrams
Evaluation and testing of layout	Bubble diagrams
Final Plan with Improvements	
Environmental impact statements	Script and maps and charts
Landscape and wildlife improvements	Script and maps
Sedimentation and erosion-control	Script and maps
Water improvements	Script and maps

LINCOLN SCHOOL GROUNDS
A Plan for
More Diversity for Better Education

The page should also have a line or two stating the address of the site. Also, a very simple, small-scale map can add decoration and express graphically the location in the state, county, or city (depending on how far away your readers may be).

A picture of the site on the title page will help create instant interest, stimulating people to look further into the proposal. The picture may be a clear black-and-white photograph with good contrast between lights and darks, or it can be a simple line drawing in black ink, drawn from a photograph or drawn at the site. A color picture is usually too expensive to reproduce, but a black-and-white line drawing can have a few dabs of color added to each copy. Color does excite interest, but remember to keep all graphics as simple as you can. Do not clutter them with unnecessary lines or words. The date on which the project is submitted or published can be near the bottom of the title page, along with your name or the name of your team or whatever group is making the proposal.

The second page logically has a table of contents listing the major sections of the proposal, with page numbers. There may be space on this page for a larger-scale location map showing the neighborhood of the proposal and the roads that lead to it.

Phase One

It is very difficult to suggest a single method of illustrating the many ideas, facts, and steps that may be involved in a project. Everyone has a personal style and some degree of artistic talent; these should be reflected in a project. The following pages point out some graphic forms that can be used. Examples of illustrations can be found throughout this book.

Illustrating the existing and to-be-constructed features of a site to be developed can be done in script (that is, written, as opposed to pictures). Any number of items might be described, from the buildings and parking spaces to lampposts and the width of sidewalks.

Maps can be used in Phase One to illustrate the need or capacity of the community to use a facility. For instance, a regional map showing movement of populations and other trends that have a regional influence on the construction of the site can be drawn. For market studies, charts and graphs can show statistics such as percentages or numbers of people, money, or other influences that warrant building the development (see chapter 6). A regional or neighborhood map can illustrate community design, existing zoning, projected zoning in the master plan, availability of utilities, or the accessibility of the site to travel routes. (See Figure 5.1.)

Maps representing natural resources of the site should attempt to summarize data collected. The graphics here can include sectional diagrams (profile views) to illustrate problems or relationships, as well as plan views of the study area. (See chapters 7, 8, 9, and 10 for examples of graphics showing studies of vegetation, soil, and water).

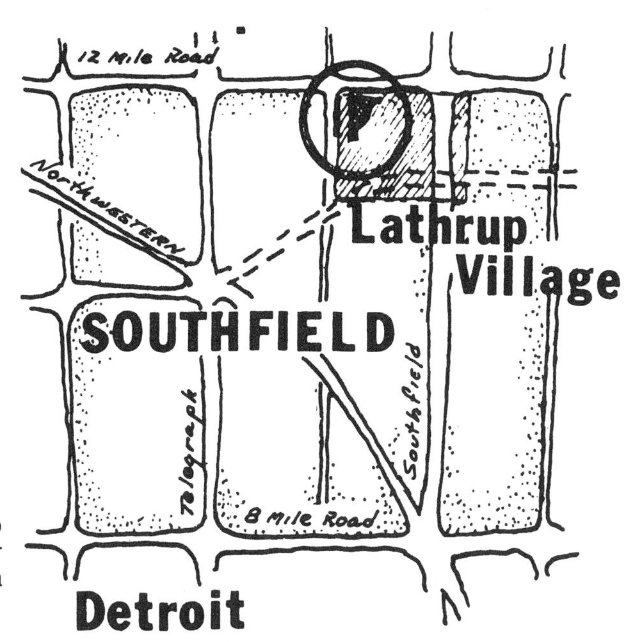

FIGURE 5.1. Regional map showing the location of major highways as they relate to a site.

Phase Two

Illustrating a framework in which a final master plan can be designed usually requires that large areas or unique features be drawn as zones of interest. As a result, these areas can be shown with *bubble diagrams*. These use circles, squares, or other quickly drawn symbols to represent an idea or area (see figure 3.3, page 31, and chapter 11 for examples).

Phase Three

In this phase of the project, the final plan is drafted (see figure 5.2 and chapter 12). Care should be taken to illustrate the proposed layout as it will appear when constructed. This final plan will be presented to the public and community decision makers, so time should be spent in making the illustrations as realistic as possible. Drawings, more realistic than diagrams, can show with three-dimensional effects the character of important features and help give the flat plan a more tangible, real, and active appearance. (See Figure 5.3.)

Often the final plan cannot describe in enough detail the features or changes that must occur. Consequently, additional written details and drawings should accompany the final plan as appendixes for those who wish to take the time to

FIGURE 5.2. Illustration of a master plan.

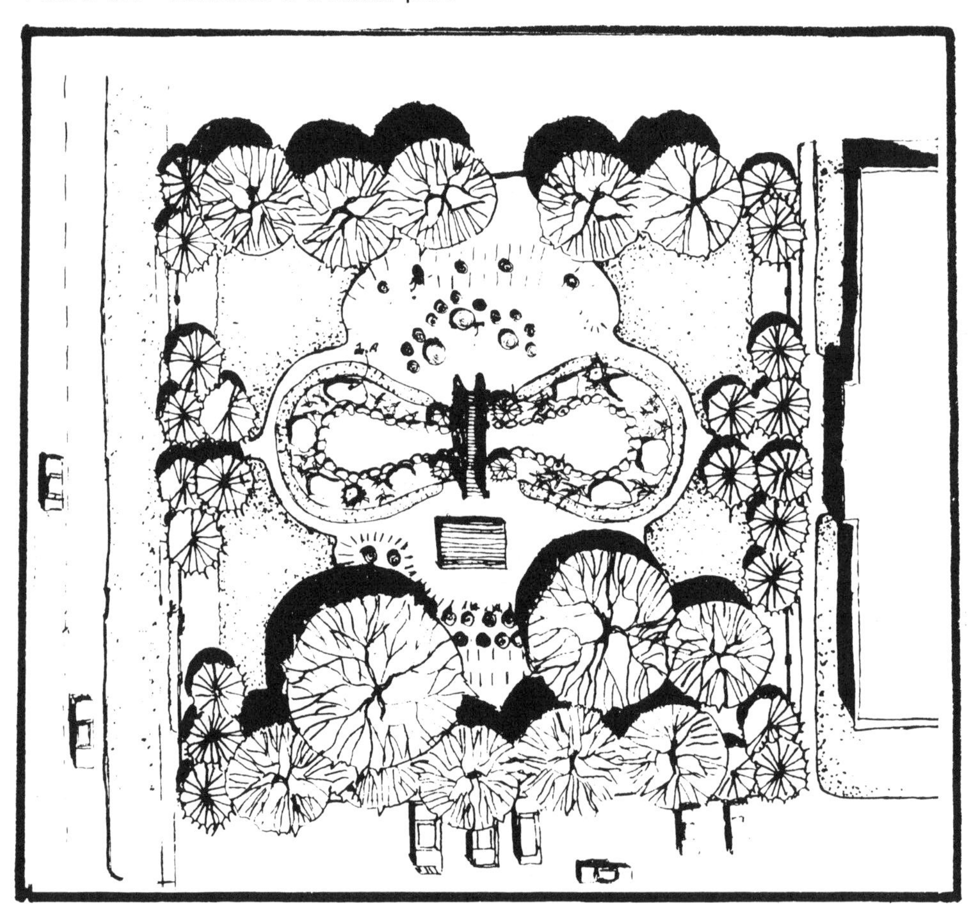

FIGURE 5.3. Character sketches may be used in a final plan to show unique features or human activity.

study them. (See Figure 5.4.) Environmental-impact studies, landscape plans, and water improvements should be as descriptive in construction details as possible. Those who must implement the plan must be able to construct the features from these drawings.

Sometimes a concept can best be illustrated by using a sectional view or block diagram of a landscape. (See Figure 5.5.) While a normal map is a plan view of a landscape, a sectional view is from the side, as if you could cut a section of land with a knife and look at it from eye level. Sectional views can be

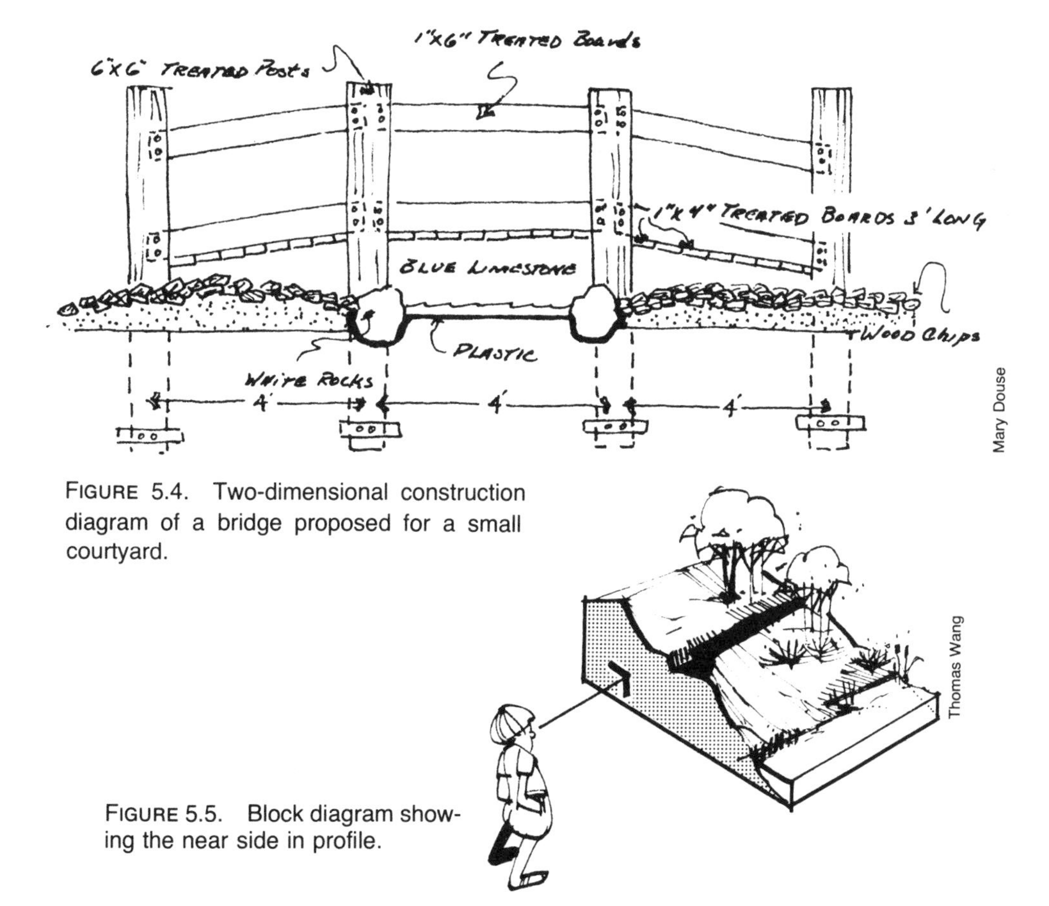

FIGURE 5.4. Two-dimensional construction diagram of a bridge proposed for a small courtyard.

FIGURE 5.5. Block diagram showing the near side in profile.

transposed from topographic maps. The problem with a sectional view is that very often the scale must be exaggerated unless the landscape is hilly. On most maps the horizontal distance is too large to make a realistic vertical profile of the same landscape in the same scale.

GRAPHIC MEDIA AND MATERIALS

Not only should a proposal be functional, but it should also be aesthetically pleasing. A proposal can be well thought out and the illustrations carefully drawn to represent one's ideas, but it will lose much of its impact if the illustrations and script are not appealing.

Artistic ability helps, but just following some basic graphic principles can make any project attractive and presentable. You should be knowledgeable about the media and materials that are available for putting proposals together, as well as aware of how to illustrate attractively. (See Figure 5.6.)

Media (plural form of *medium*) are the kinds of materials used in illustrating: for example, pencils, charcoal, chalk, pen and ink, crayons, felt-tipped pens, water color paints, acrylic paints, and oil paints. The term *media* may also refer to other materials such as felt and flannel boards, films, and books. A particular medium, for instance photography, may include many different techniques, such as still photography with either prints or transparent slides for projection, or motion picture photography of many sorts. (*Graphic* means presented visually on a relatively flat surface like paper or chalkboard, two-dimensionally. *Glyphic* is the corresponding term for three-dimensional presentations such as carving, sculpting, and modeling.)

Generally, your audience will determine the medium you should use. Obviously, the plans you would make for constructing a home for your family would be different from those required for constructing a shopping center. In the first case, a construction blueprint might be the only drawing necessary. In the case of the shopping center, many people would have to be involved in the decision-making process and more illustrations would be required.

In schools, most work can be done on standard 8½″ × 11″ bond paper of a

FIGURE 5.6. A somewhat diagrammatic profile sketch of a proposed planting.

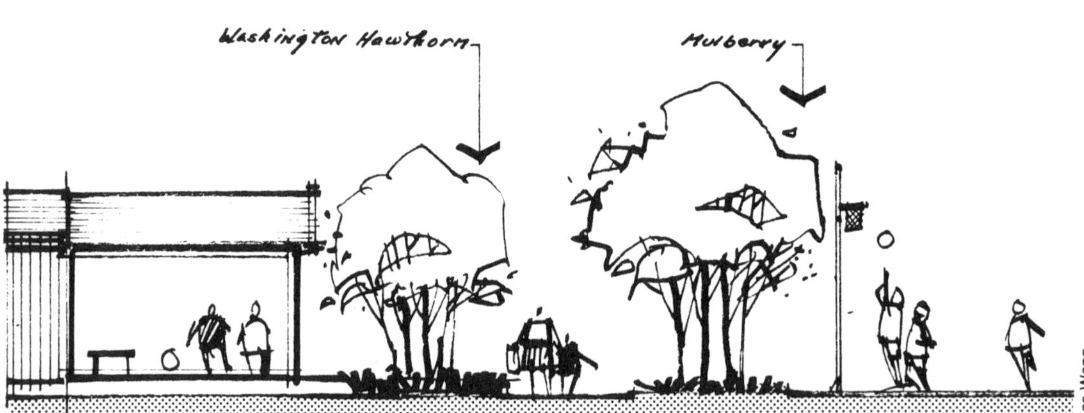

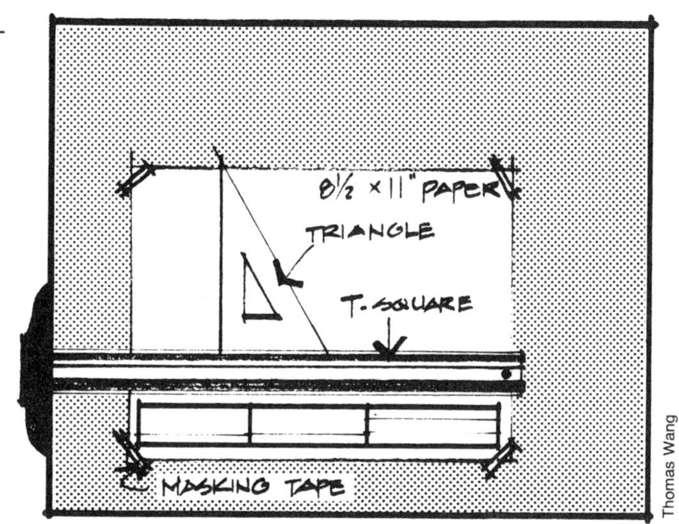

FIGURE 5.7. Sketch of student's desk top.

kind usually available in school. Most schools have equipment for making reproductions on this size paper, as well as photographic equipment and the equipment necessary for making transparencies of the finished products. Inexpensive drafting and drawing materials are also available in most schools. The classroom desk can act as a drawing table. Small T-squares can be purchased that make the job of drawing straight lines easy; small plastic triangles are available for drawing vertical lines. Masking tape can be used to secure the paper to the desktop. (See Figure 5.7.)

Some people recommend hard pencils (4h–6h) to sketch in the rough outlines and details of an illustration. Sketch lines should be made as light as possible so they will not show when the final lines are added. Others recommend a softer pencil (such as No. 2 writing pencil), kept sharp; its lines are easily erased until barely visible prior to making the finished lines with black ink.

Along with paper, most schools or office facilities have India ink, soft graphite (No. 2) pencils, and copy machines available to teachers and students. India ink reproduces best for maps because of the amount of carbon in the ink. It produces sharp, clear prints. If blueprints are desired, opaque paper must be used. Professional drafters often use transparent plastic from which many copies can be made.

Today, top-quality ink pens that do not leak and that carry a large amount of ink can be purchased. The more old-fashioned steel pen-points that can be set in a wooden or plastic handle are much less expensive. Ruling pens available at art supply stores make neat lines for such jobs as drafting borders and scales on maps, and making diagrams. Correction fluids that will cover mistakes and not show up on the reproduced copies are available. Colored pencils, pastels, and felt-tipped pens can be used to add color to illustrations.

MAKING YOUR ILLUSTRATIONS

You should follow certain basic design criteria in illustrating a concept or idea. As a rule, an illustration may have room for the drawing, a caption describing it, and a legend with such standard information as the name of the designer,

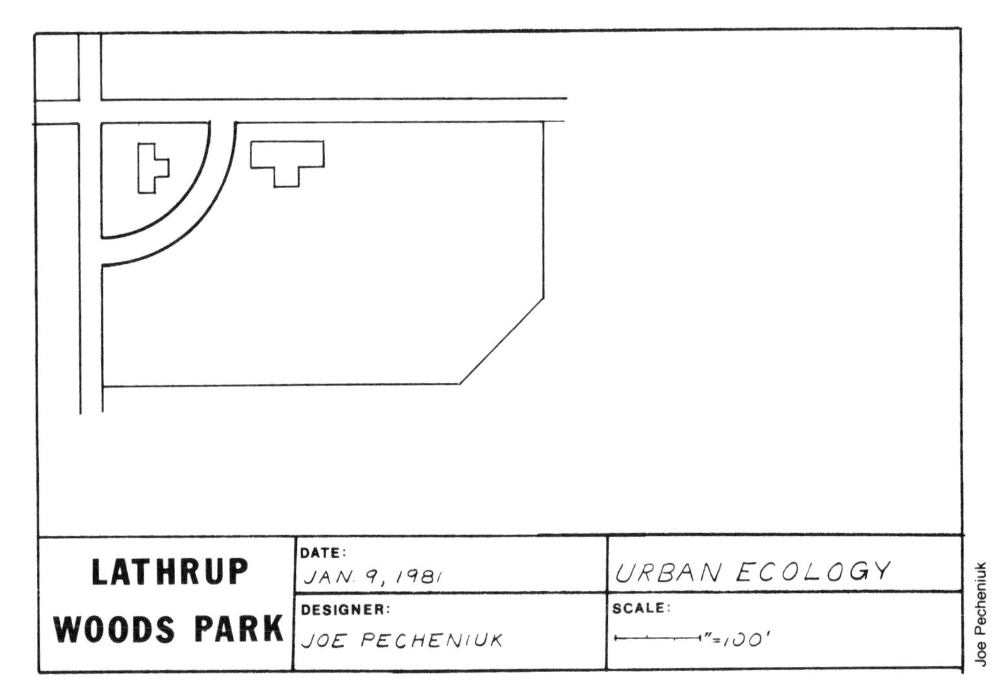

FIGURE 5.8. Sample base map.

date, and the name of the project. (A map will of course include also the information described in chapter 4.)

The first step in laying out an illustration is to decide where the drawing, caption, and legend should go. Several principles of *spatial organization* should be followed here. The first is to allow for a border around the entire illustration. This will serve to frame the illustration much as a frame does a picture.

The same principle should be used in separating the areas of content within the illustration. Study newspapers, magazines, and textbooks for good examples of how the content of a page can be spaced in an orderly fashion, with plenty of empty space between items and the items lined up neatly.

For many illustrations such as maps showing soil, vegetation, topography, and water, a standard *base map* can be constructed and used as a master. (See Figure 5.8.) A base map is a map of the site with relatively permanent existing features such as rock outcrops, landmark trees, roads, trails, and buildings. Once a base map is drawn, numerous copies can be reproduced to serve as a base for recording site notes or summarizing these notes into resource maps and master plans. Make the base map as simple as possible to provide room for specialized data to be added, keeping northpoint, scale, and date small, usually in a corner.

Basic site features can be obtained from updated aerial photographs, topographical maps, zoning maps, and so forth. Many accurate maps can be obtained from the county unit of government where property descriptions are recorded (see chapter 4).

Transparent slides and opaque photographs can help show the character of a site. Pictures of your site can be taken and used as aids in describing the area, or sketches can be made from pictures by tracing them on a *light-table* or against a bright window, or by projecting the picture on drawing paper fastened to a wall.

If you want to enlarge your map, the opaque projector found in most schools does an adequate job, or an overhead projector can be used. The area of the map enlarged must first be reproduced on transparent paper if an overhead projector is used. This can be accomplished on a copying machine using light-sensitive transparency paper, or by tracing the lines onto a plain sheet of matte plastic.

If a map of the same size is required, simply trace the map you wish on a light-table or window. Enough light is available from a window to supply you with a basic sketch of your map; it can be more carefully redrawn after you take it down. Tape the original map to the window, then tape a sheet of white paper over the map. With a carbon pencil or pen, trace the features that you think are necessary. Once a map has been drawn, a certain number of words are necessary to describe its contents. Sometimes one or two words suffice to describe an item on the map. Sometimes you will have to use several sentences, or even a number of paragraphs in your description. In general, it is best to *use as few words as possible*; let the drawing speak for you. (See Figure 5.9.)

If labeling is necessary, try to give the labels different styles of lettering or "value strength." On any map certain ideas are more important than others and should stand out more. The title of the map, for example, might be made with a bold style, but should not take too much space.

FIGURE 5.9. Vegetation map copied from an aerial photograph.

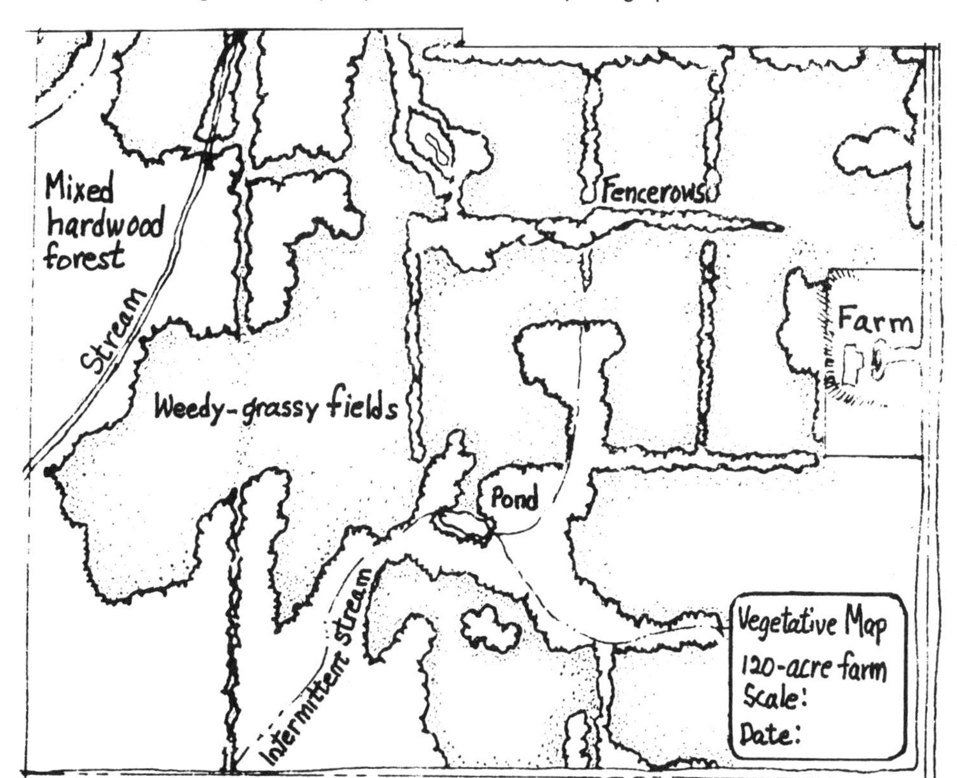

Sometimes it may be desirable to make a drawing more attractive. Color will help. Practice with different colors and combinations. Practice on tracing paper or clear acetate placed over a map. Be sure the colors you use will appeal to your audience. It may help to give value strengths (lightness and darkness) to the colors. For example, dark brown can be used to represent poorly drained muck soils that are bad to build upon. Light brown can reflect sandy soils suitable for drainage fields from septic tanks.

Finally, make all the lines, colors, and so forth that go into a drawing sharp and clear and stay with the same style throughout the project, unless you have a good reason not to do so (as in this book, which strives to illustrate a variety of styles).

CONCLUSION

The format and techniques of illustration presented in this chapter are meant only to serve as a basic approach to working with land-use issues. Plans will vary depending on the developer's needs, the nature of the site, the impact the proposed changes may have on the community, and the planner's ability to communicate the ideas and data involved in the study. Whatever the problem, a systematic approach and careful illustration of your ideas will improve the chances that your plan will be implemented.

REFERENCES

Cayton, George T. *The Site Plan*. Champaign, Ill.: Stipes, n.d.

Ching, Frank. *Architectural Graphics*. New York: Van Nostrand Reinhold, 1975.

Daniels, Alfred. *Drawing For Fun*. New York: Doubleday, 1975.

Edwards, Betty. *Drawing on the Right Side of the Brain*. Los Angeles: Tarcher, 1979.

Gollwitzer, Gerhard. *Express Yourself in Drawing*. New York: Sterling, 1967.

Guptill, Arthur L. *Freehand Drawing Self-Taught*. New York: Watson-Guptill, 1980.

Hartmann, Robert. *Graphics for Designers*. Ames, Iowa: Iowa State University Press, 1976.

Jameson, Kenneth. *You Can Draw*. New York: Watson-Guptill, 1976.

Kemmerich, Carl. *Graphic Details for Architects*. New York: Praeger, 1968.

Lutz, E. G. *Drawing Made Easy*. New York: Scribner's, 1921.

Pitz, Henry C. *How to Draw Trees*. New York: Watson-Guptill, 1972.

Sheaks, Barclay. *Drawing and Painting the Natural Environment*. Worcester, Mass.: Davis, 1974.

Walker, Theodore. *Plan Graphics*. West Lafayette, Ind.: P.D.A., 1975.

Wang, Thomas C. *Plan and Section Drawing*. New York: Van Nostrand Reinhold, 1977.

Watson, Ernest W. *Ernest W. Watson's Course in Pencil Sketching*. New York: Van Nostrand Reinhold, 1967.

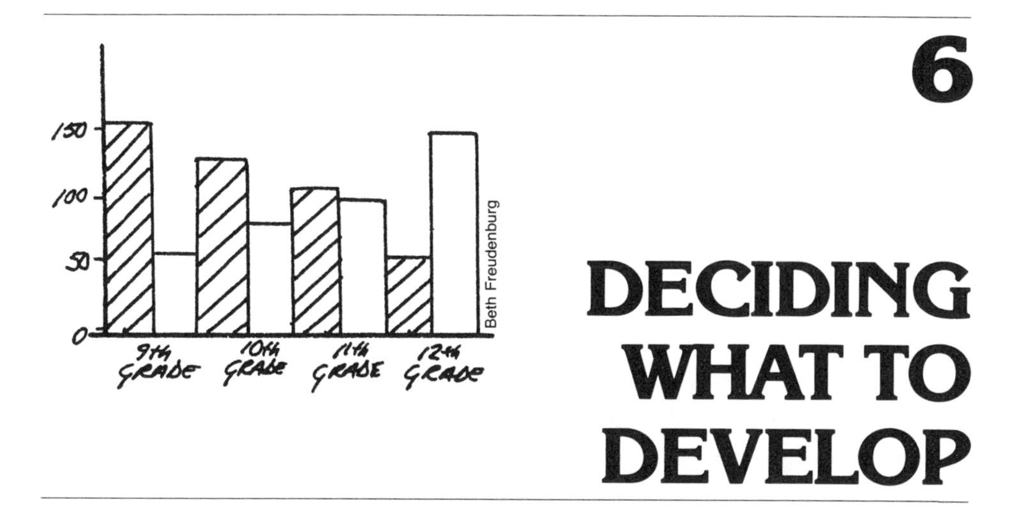

6 DECIDING WHAT TO DEVELOP

Until now we have taken a general look at planning. We have studied problems inherent in the existing planning process and problems that are the result of urbanization. We have looked at one alternative, an ecological planning process, and in chapters 4 and 5 we examined methods of collecting and illustrating site data and ideas.

In this chapter we take a more direct look at the ecological planning process discussed in chapter 3. Here we are primarily interested in market studies (studies that can help determine whether a proposal will be successful or rewarding to the developer) and community-influence studies (studies that can help determine whether the development will fit into standards established by the local community).

MARKET STUDIES

Let's pretend that you own a piece of property and you want to know what type of development will generate public interest. Making a decision about what to build may seem like a simple task, but a hasty decision can often lead to disaster. To help make an intelligent decision about your development, you may wish to do a *market study*. Market studies use population statistics to identify population trends or shifts that may have an impact on a proposed development.

A *population* can be defined as the total number of inhabitants in an area, such as a classroom, city, county, and so forth. *Population statistics* is the science that deals with the collection, classification, and use of numerical facts or other data that pertain to a population. Population statistics may include birth rates, death rates, and growth rates. Population statistics may also include data about the age of the people in a population, their habits, likes, and dislikes. Obviously the data collected on a given population can be endless unless the planner can identify and isolate the data that pertain to the project. The planner must ponder such questions as, Who are the people that constitute the population that will use my facility? What is the exact or approximate size of the popula-

FIGURE 6.1. Different people have different values often indicated by their life styles and architecture.

Photos by John Brainerd

tion? What statistics about this population do I need to identify to help me make an intelligent decision about my project?

Sources of Statistical Data

Market studies usually require the expertise of market specialists or *demographers* (experts in population statistics). In the past, market studies were seldom carried out. One of the marks of good businessmen or administrators was their ability to "feel" what the market might be for a proposal. Today, many private marketing and planning firms specialize in market studies; large businesses utilize their services. These private firms have access to large volumes of government statistics and conduct surveys to collect the data themselves. They also analyze the data and help make decisions about the feasibility of a proposal.

In addition to private firms, many public agencies can supply demographic information to the planner. The U.S. Department of Commerce and its Bureau of the Census and the U.S. Department of Labor and its Bureau of Labor Statistics are two agencies interested in population, employment, and housing. Census reports and other publications from these agencies can be obtained from the Superintendent of Documents, U.S. Government Printing Office, Washington, D.C. 20242.

County government can be helpful in supplying data and maps pertaining to ownership of land. If your county has a planning department, it can also be very helpful with information on trends or shifts in population and how various units of government are planning for these population shifts.

Local banks, utility companies, planning commissions, title insurance companies (in some states), zoning boards, and building inspectors are good sources of local statistics. Usually the regional planning agencies that have been formed to help plan and coordinate regional programs are a rich source of data. Ask for their names and addresses at your town or city hall.

Local colleges and universities are another source of information. Schools or departments of business, landscape architects, and urban and regional planners are just a few sources of assistance.

Collecting Statistical Data

In some cases, statistics are not available or are inadequate for making decisions. Where such a situation exists is may be necessary to collect data on your own. Two methods used for obtaining statistical data on populations are the *true census* and the *sampling estimate*. The true census is a count of all the individuals (or their feelings, habits, and so forth); it is the most accurate way of collecting information. On small projects around the school this method can prove successful. Unfortunately, this method is not practical in most circumstances because of the time and expense involved in surveying large populations; therefore information about large populations is usually obtained from sampling estimates. These are obtained by collecting information from portions (samples) of populations which, if collected properly, are assumed to be representative of the total population. Accuracy depends on the sample being collected at random, on the validity of the instrument used to collect the information, and on the investigator's ability to interpret the samples.

The techniques used to gather the information may consist of constructing a *questionnaire*, using a *checklist*, or conducting a *verbal survey*. A questionnaire

RESULTS OF PARK SURVEY OF 1,000 STUDENTS AT LATHRUP HIGH

	PASSIVE PARK	ACTIVE PARK
SENIORS	750	250
JUNIORS	630	370
SOPHOMORES	520	480
FRESHMEN	260	740

Beth Freudenburg

FIGURE 6.2. Table of park survey data.

is a paper-and-pencil instrument on which an individual is asked to respond to a series of questions. The questions are designed to find out the individual's preference toward the proposed project or issues related to the project. Checklists are instruments designed by the investigator to rate responses of individuals, usually by observing response patterns. Verbal surveys have the same function as questionnaires and checklists but are carried out on a one-to-one basis over the telephone or by personal contact.

Illustrating Statistical Data

Maps, charts, graphs, and tables are the commonest methods of presenting statistical data. Tables have the least impact on an audience, but they may be the most objective. The information in tables is usually listed along with a caption that may briefly describe and interpret the data. Because the raw data are present, readers are often left to reach their own conclusions about their meaning.

For example, the results of a student survey on a park might be shown in a table. (See Figure 6.2.) Students were asked to choose between an active park, where such activities as tennis and Frisbee could be played, and a more passive park, where eating, reading, discussion, and classroom activities could take

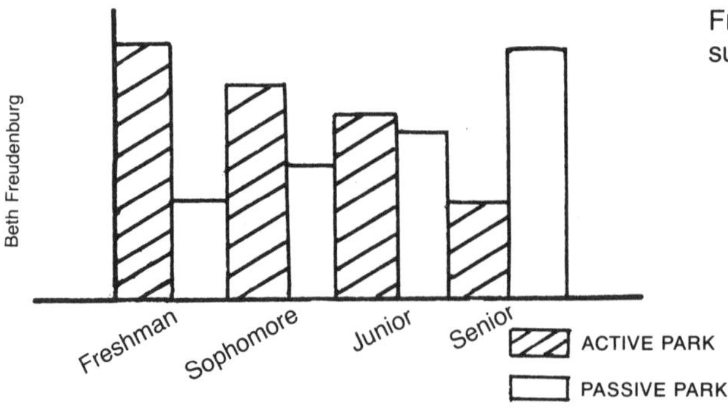

FIGURE 6.3. Bar graph of park survey data.

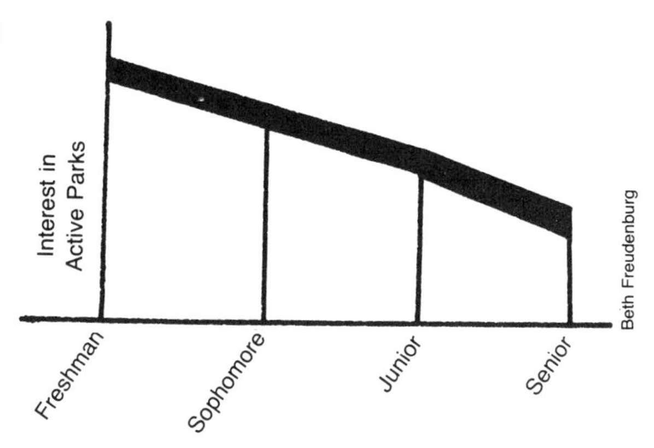

FIGURE 6.4. Line graph of park survey data.

place. The results of the survey are shown in the table simply by number of responses per grade level.

These same data could be put in a graph or chart form, which might make the information more appealing and easier to understand and interpret. (See Figure 6.3.) The same information can be presented with lines. Figure 6.4 shows that active parks are more appealing to underclassmen. You could draw the same kind of graph for passive parks.

Maps can also be useful tools in illustrating market studies and regional or neighborhood influences. Maps or plastic transparencies laid over a base map can effectively show the trends or shifts in populations. Factors involved in such trends, such as the movement of populations along major transportation routes, can be illustrated with arrows and different kinds of shaded patterns. (See Figure 6.5.)

Analyzing Statistical Data

Throughout the planning process there is a need for making value judgements and decisions about an issue. Statistics produced by a market study require

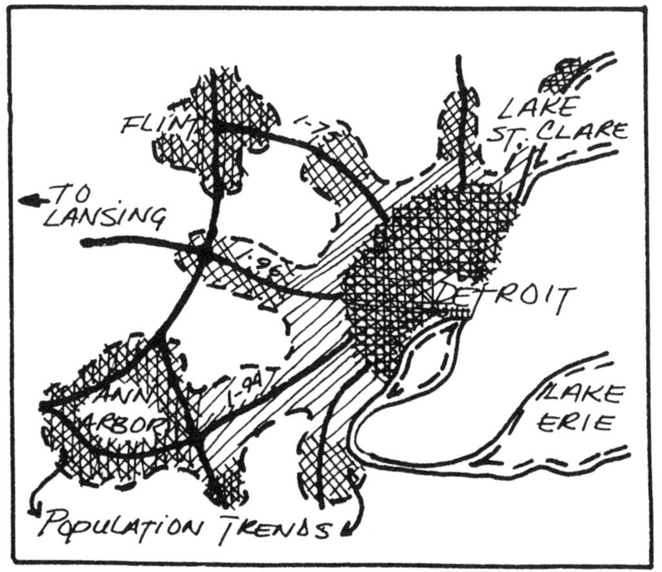

FIGURE 6.5. Map of population trends.

Barbara Brainerd

FIGURE 6.6. Living in a rural area may mean driving a long distance to work. Costs and benefits must always be considered and trade-offs made.

analysis just as surely as do data from site studies showing types of soils, vegetation, water communities, and wildlife habitats. It is the planner's job to analyze these data and make recommendations relative to their impact on the proposed development. Site studies and market studies only supply data from which intelligent decisions must be made.

You must remember that there is seldom a right or wrong solution to a problem: there are only different solutions, some of which are better under certain circumstances. Costs and benefits must always be considered and trade-offs made. Living in a rural area may mean driving to work every day, often over poor roads, but you have more outdoor living space. That is a trade-off.

A number of systems can be used to help make decisions. These systems range from the most sophisticated computerized systems-analysis programs to the old "muddling through" processes. Though most analytical systems are complicated, the data collected in the average classroom will lend themselves to a simple technique called *value analysis*. Value analysis involves placing a number value on the opinions you, or your group, have on an issue.

The "Value Analysis Form" shown in Figure 6.7 is an example of a value analysis chart that can help you state your opinions on an issue and assign numerical values to each statement. The use of a value analysis chart helps one to examine an issue systematically and somewhat objectively arrive at a consensus about it.

The basic idea of value analysis can be enhanced and improved upon by including a second step. Members of a group can determine their own individual conclusions with the individual analysis chart and then submit the charts to the group for group discussion and critiques. A group concensus can then be arrived at using the same technique.

Proper analysis of population statistics can also help answer such questions as: Will the existing demand for apartment housing continue? Will the commu-

Name June Covil Date 11/12/79

Land-Use Issue Fast-Food Restaurant near Our High School

	POSITIVE VIEWPOINTS	VALUE		NEGATIVE VIEWPOINTS	VALUE
1.	Create jobs	10+	1.	Traffic congestion	7-
2.	Offer variety of food	9+	2.	Attendance problems	5-
3.	Place to socialize	5+	3.	Litter problems	9-
4.	Place to relax	2+	4.	Noise pollution	2-
5.			5.	Visual pollution	1-
6.			6.		
7.			7.		
	Total	26+		Total	24-

Conclusion: I am slightly for the fast-food restaurant.

FIGURE 6.7. Value analysis form.

nity need a new junior high school? In the first case one must know the rate at which the population is growing; in the latter case, one must know the rate and direction of growth in the school population.

The growth rate of population can be calculated if the population of a community is compared to the number of individuals entering and leaving the population or community. For example, the growth rate of a community of 12,000 can be determined if data on the number of births, deaths, immigrants, and emigrants are available. These data can then be incorporated in the following formula:

$$\text{Community growth rate} = \frac{(\text{Number of births} - \text{Number of deaths}) + (\text{Number of immigrants} - \text{Number of emigrants})}{\text{Total population}} \times 100$$

By modifying the formula slightly, other population growth rates can be calculated. The rate of growth of your school can be determined if the population at the beginning of the school year is known and data on the number of students entering and leaving the school during the school year are available. The following formula illustrates such a situation:

$$\text{School growth rate} = \frac{(\text{Number of students entering} - \text{Number of students leaving})}{\text{Total school population}} \times 100$$

$$\text{School growth rate} = \frac{50 \text{ entering} - 10 \text{ leaving}}{1{,}000 \text{ school population}} \times 100$$

$$\text{School growth rate} = \frac{+40}{1{,}000} \times 100 = 4\%$$

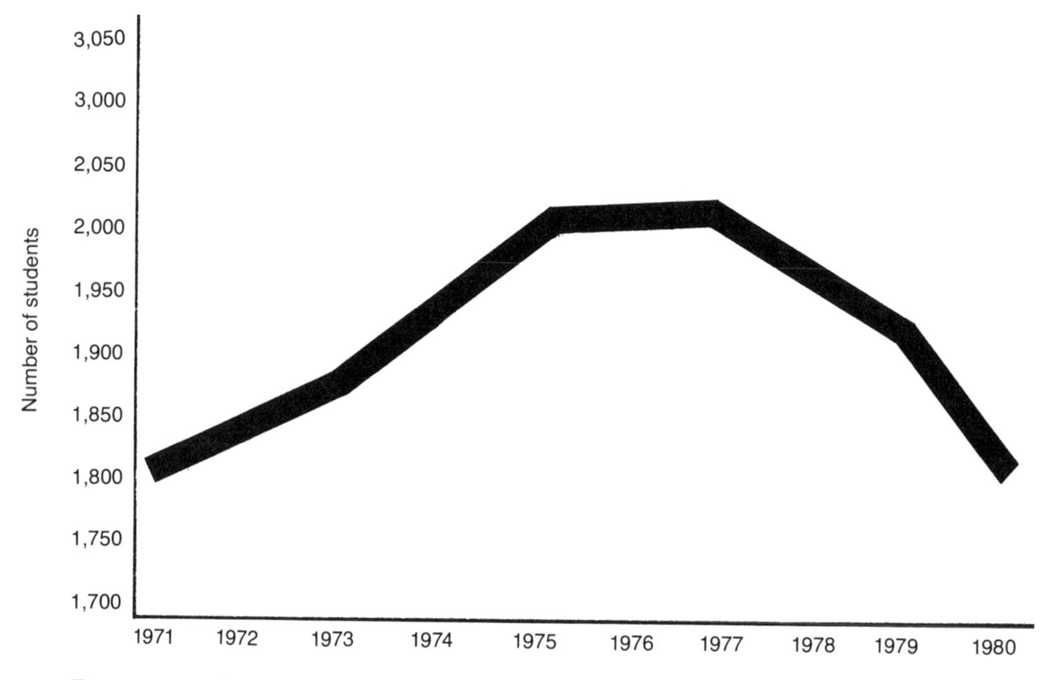

FIGURE 6.8. Graph showing trends in a high school population.

Plotting population statistics over a time period can also be helpful in projecting trends. Figure 6.8 shows the growth rates for a high school over a ten-year period. Several trends are evident. The period from 1971 to 1975 shows a rapid growth, whereas the period from 1977 to 1980 illustrates a decline in school population. Trends may suggest that action be taken to reduce school construction, hire fewer teachers, and enact or recommend other measures such as closing some schools to meet the reduction in school population.

A market study may also explore the nature and condition of existing facilities in the area that would be in competition for the same market. For example, one barber shop per ten square miles may be the accepted ratio for barber shops. Thus, if a survey of existing facilities reveals that this ratio already exists in the proposed area, it may be advisable to look for another area in which to open a shop.

All the preceding types of studies imply more than just collecting data about a given population. They suggest that projections be made relative to the needs or interest of the population. A population of college students will probably have a different set of interests and needs than a younger or older population. A new subdivision of three-bedroom homes can be expected to generate young married couples, who, in turn, will create a market for babies' needs. Neighborhood stores might stock baby items for a period of time, then shift their emphasis to bubble gum and baseball cards as the population grows older.

COMMUNITY INFLUENCES

Phase One of the ecological planning project also suggested a community influence study. Once the nature of the development has been determined, the planner should study the surrounding area carefully for relationships that may be beneficial or harmful to the proposed development. Most urban and rural communities have an established character they wish to preserve, sets of legal standards with which developments must comply, and basic services that are available to any new development. These factors should be studied carefully, because most legal conflicts start here.

Community Services

Accessibility of the site to the people it will serve depends on the existence and condition of community roads or other means of transportation. As a rule, a site developed on a good road has a better chance of success than one on a poor road. Other factors are also involved, such as the speed limit on the road and the location of stop signs near a business. The fact that most gasoline stations are located on corners with stop signs is a good example of what a sign can do.

Other community services should also be nearby. In some areas, water can be obtained from underground sources if community water is not available. In some states, permission of the county is required to dig wells. In addition, septic tanks for handling sewage are adequate in areas where the population density is low and the soil suitably porous, but city sewage facilities must be available in more concentrated urban areas. Gas, electricity, and telephone are other services that must almost always be available, and their utility lines located and shown on maps of a proposed development.

FIGURE 6.9. Factors such as the location of a stop sign can be important in the success of a development.

John Brainerd

Legal Regulations

Legal standards can be imposed upon most developments by the local, county, state, and federal governments. Zoning ordinances and building codes set standards that planners must consider. *Allowable density* (number of dwelling units per acre), *height restrictions*, and *set-backs* are the most important. Height restrictions set limitations on the number of stories a building can have. Set-back regulations govern the number of feet a building must be set back from a neighbor's property line or a road.

The location of *easements* and other legal *rights-of-way* should also be noted. Easements and rights-of-way give legal access to your property for the purpose of public welfare. These easements are usually obtained by government agencies or public utilities, which may use portions of your land for transporting their product from one place to another, for instance running a water main.

Easements for gas and electricity are among the most common, although some communities today are experimenting with easements for such purposes as recreation. The legal edge of roads often extend beyond their obvious limits. These road rights-of-way are established in the event the road has to be widened; they are usually measured from a surveyed center line of the road.

Peripheral Influences

Many legal problems result from a development's inability to fit the established character of a community or neighborhood. Property owners may attempt to develop a property according to their own needs and desires, the nature of which may be of concern to neighbors. Painting your house bright purple might be an extreme example of rejecting neighborhood standards. A tavern built next to a church or school could cause conflicts. When an incompatible land use is granted approval within the confines of a zoned district, it is called "spot-zoning." Most planning commissions try not to spot-zone, so it is usually difficult to get a variance for a nonconforming land use.

CONCLUSION

Clearly, there are many factors to consider in proposing a development. Some facts are best collected statistically. Some, such as people's attitudes, must be felt as well as counted. Laws, developed according to society's wishes in the past, must be respected. Your attention to such matters will be reflected in both the writing and the illustration of your proposal.

You will do well to make several "roughs" (rough outlines, sketches, and maps) before settling on a pattern of land-use change. Many sheets of scratch paper may be used up before you say, "Now I think I have it!"

In time, you will develop your own planning techniques. One excellent way that saves time and paper is to cut out pieces of thin cardboard to represent any building, parking lot, tennis court, septic tank and drainage field, or other area. These can be moved about on your base map until they seem to fit comfortably. (The same technique works well when planning living room furnishings: it is easier to move cardboard several times than to shift sofas and pianos.) When you think ecologically, you realize that the pieces must fit both the natural resources of the landscape and the needs of society.

REFERENCES

Association of American Geographers. *Geography in an Urban Age*. Toronto: Macmillan, 1970.

Chinitz, Benjamin, ed. *City and Suburb*. Englewood Cliffs, N.J.: Prentice-Hall, 1964.

The City Game. Anaheim, Calif.: Dynamic Design, 1970.

Ehrlich, Paul R., and Ehrlich, Anne H. *Population Resources Environment: Issues in Human Ecology*. San Francisco: W. H. Freeman, 1972.

Investigating the Human Environment: Land Use. Boulder, Colo.: Biological Science Curriculum Study, 1979.

McCue, G. M. *Creating the Human Environment: Report of the American Institute of Architects*. Urbana, Ill.: University of Illinois Press, 1970.

McEvedy, Colin, and Jones, Richard. *Atlas of World Population History*. New York: Penguin, 1978.

McKeever, J. R. *The Community Builders Handbook*. Washington, D.C.: Urban Land Institute, 1968.

Thomas Wang

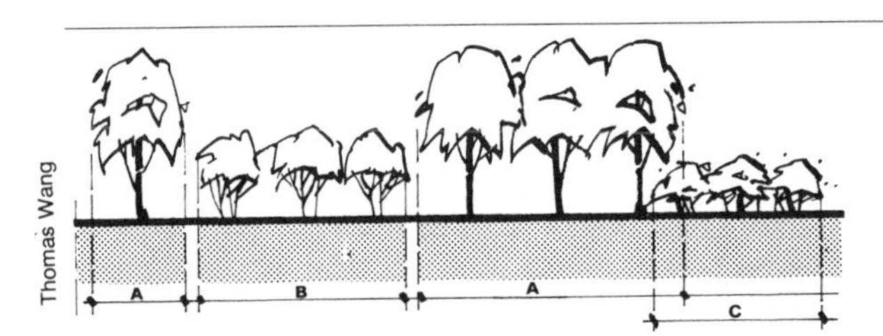

7 INVENTORY AND ANALYSIS OF PLANT COMMUNITIES

In this chapter, we emphasize the importance of plants in our environment, so we talk about certain kinds of plants living together in *plant communities*. In so doing, we must not forget that what we are really dealing with are *plant-animal communities*, more often called *biotic communities*, because plants and animals do not live independently of each other. And because the plants and animals are also an integral part of the physical environment of air, rock, soil, and water, we must think of plants as part of very complex systems called *ecosystems*. In ecological planning we, of course, cannot deal with only one factor at a time, nor can we deal with everything at once. So in this chapter, we focus on plants living together in plant communities but try to keep in the back of our minds the complex systems of which they are a part.

Plants play an important role in our lives. Directly, through the vegetable tissues they produce, and indirectly, through the proteins, carbohydrates, and fats of animals that eat them, plants supply us with vital nutrients. They are also an important part of our visual environment. As with the air we breathe, we often take the aesthetic value of plants for granted, never fully realizing the effects they have on our lives. But just stop to think what a dull world it would be without plants in our landscape.

Living in a rural environment, people were close to the plants, cultivating them for food, using them to heat homes, and harvesting the wildlife that used the plant communities as its home. But our style of life has changed. Life is more frantic today in our bustling cities. As our life style has changed, so have our attitudes toward plants. Except in small pockets of usually expensive real estate, plants and their varied functions are often slighted by environmental designers and many citizens.

We are beginning to realize again that plants can help provide us with a positive, natural solution to many environmental problems. Plants function as air purifiers and insulators; as buffers against heat, noise, and light; and as screens to minimize unpleasant views. Plants also help control microclimates in our urban areas and combat the effects of erosion and sedimentation problems caused by the stripping of vegetative cover from the surface of the soil.

The profession of landscape architecture has evolved around the skillful

FIGURE 7.1. Plants help make our environment pleasant.

Dave Larwa

placement of plants into the landscape. Traditionally, landscape architects were primarily concerned with plants and their arrangement, but fortunately many landscape architects have expanded their practice to include assessment of community needs and inventory of community and site resources when preparing and designing master plans.

In this chapter, we shall be concerned with conducting a vegetative and wildlife survey of a site to be developed, for the following reasons:

1. To identify the location, size, and configuration of any plant communities that should be considered when developing a scheme for laying out developmental units
2. To help estimate the types and relative abundance of wildlife on the site
3. To identify any unique or rare plants and animals
4. To help estimate the size and types of soil deposits, valuable minerals, and archeological features at the site

ECOLOGY OF PLANT COMMUNITIES

The planner must be familiar with the many ways plants function within natural and man-made environments. In the natural environment, individual plants are associated with plant communities. To a large degree, the plants within a plant community are subject to the soil and atmospheric conditions that prevail in that area. This principle is true even on a small scale (e.g., a crack in the sidewalk may be considered a microcommunity). As a result, where soil and water conditions vary, so do the plants and animals associated with these environmental conditions.

From one point of view, plants should be studied after soils and water, for without the latter two there would be no plant life. But plants as a group are excellent indicators of air, soil, and water relationships. As a result, it can be advantageous for planners to conduct a vegetative study early in the planning process, thus letting the plants tell them about the microclimates, soils and water.

For example, in the wet, organic soil of a tamarack bog, only species of plants that have adapted to the acidic conditions can survive. Plants growing on a forest floor must be able to tolerate low temperatures, little sunlight, and moist soils. A short distance away on a barren soil, a desertlike condition may exist. In this more severe microclimate, only hardy plants like chicory and dandelion can survive.

All plant communities play an important role in the natural water cycle and can be of assistance to any development in controlling erosion, cleaning water, and functioning in a variety of other ways beneficial to a landscape. The roots and leaves of trees and other plants help prevent the soil from being washed

FIGURE 7.2. Succession will occur if a field is left undisturbed.

away. Dead leaves and branches help supply nutrients and structure for the soil. A well-structured soil can control the temperature and rate of insoak and seepage and thereby provide a constant supply of cool water throughout the year.

Most plant-animal communities are not stable; that is, they do not remain the same but are constantly evolving or devolving. One type of community succeeds another until a final, stable (climax) community may be reached. This process of change is known as *succession*. For example, a plant community that represents an early stage in succession may be a farmer's plowed field. Left unplanted and undisturbed, the land will be invaded by plants capable of living in the hot, dry environment. Their presence will alter the physical conditions of moisture and temperature so that first annual weeds, then perennial grasses, then shrubs and sun-loving trees can live there.

Sun-loving trees, in their turn, create conditions that are unfavorable for their own kind but favorable for shade-loving plants to survive. Somewhat similar series of events occur in wet soil and in bodies of water. The plant communities that result will be different, but the process parallels that in an abandoned field.

Some species of plants can be found in a variety of plant communities. Dandelion is a good example of a plant that can withstand a variety of physical conditions. Some animals, too, range over a variety of plant communities. Usually, the larger the animal the greater its range, but it will prefer particular communities within that range. Don't we all?

SOME TYPES OF PLANT COMMUNITIES

Many planners are increasinly sensitive to natural landscapes and how they are being altered. At one time the United States was covered primarily with forest and prairie. Most of the original plant communities have been modified in some fashion, but because of climatic conditions the grasslands dominate throughout the Midwest, with a variety of forest in the East and patches of desert, grasslands, shrublands, and forest to the West.

Approximately 32 percent of the United States was originally classified as grassland. Tall-grass prairie communities extended through Manitoba south into central Oklahoma and eastward into Ohio and southern Michigan. Semidesert grassland extended from central and southwestern Texas into Arizona. Forest-type grasses and cone-bearing trees were dominant on the higher plains, plateaus, and mountains from the Midwest to California.

When the United States was first settled by Europeans, approximately 47 percent of the land was in forest and woodland. The process of clearing the land for agriculture and forest products reduced much of this land to farms, secondary growth, and grassland. Here again, because of soil and climate, a variety of forest communities has evolved. The Northern Coniferous Forest covers much of northern New York, northern portions of Minnesota, Wisconsin, and Michigan, and most of Canada beyond the plains provinces; spruces, firs, and associated species predominate in this forest. The Appalachian Oak Forest is located in nineteen states in the central-eastern half of the United States. The forested portion of this region occupies approximately 20 percent of the total area, with deciduous trees dominant.

Southern forests from Virginia to Texas are characterized by valuable stands of pine, bald cypress, and bottomland hardwoods. The Rocky Mountain Forest extends from Canada to Mexico and contains a mixture of coniferous trees, with ponderosa pine, Colorado blue spruce, and Douglas fir common. The heaviest stands of forest occur from Washington to northern California and are sometimes called the Pacific Coast Forest. Redwood and the giant sequoia are well-known species of this association, along with Douglas fir and Sitka spruce.

Within each of these large associations, smaller plant communities reflect unique ecological conditions or successional changes. Because planning is often carried out on a local level, planners should be familiar with small as well as the large plant communities. Some of these small communities, such as coastal estuaries, saltmarshes in the Northeast, and mangrove swamps in the South, were mentioned in chapter 2. A description of these communities is beyond the scope of this text, but a brief description of some plants and animals found in plant communities of the Appalachian Oak Forest follows to give an example of the information available.

Upland Plant-Animal Communities

Upland communities are areas of land that are well drained because of the types of soils present or the nature of the slopes. Soil composition depends on several things: the structure and texture of the soil; how long the land may have been left untouched by people; and the nature of any human activities carried out on the land. Generally, the development potential for such land is great because of its drainage and structure-bearing capabilities.

DENUDED-ERODABLE LAND

Since surface soil has been removed for minerals or other soil resources (e.g., topsoil, farming) a harsh environment is created for plants. Exposed, relatively

FIGURE 7.3. Denuded-erodable land.

John Victory

FIGURE 7.4. Abandoned field or dry meadow.

infertile subsoil is largely visible, with gully erosion or the less-visible wind and sheet erosion resulting in more loss of topsoil.

Surface soil removal may also create water problems in terms of either water-table exposure or exposure of subsurface soil with poor drainage capabilities.

ABANDONED FIELDS OR DRY MEADOWS

These well-drained, sunlit fields have soils protected by herbaceous plants such as grasses and forbs (herbs that are not grasslike). The dominant forms may vary greatly, depending on the physical conditions, as well as the sources of seeds. Recently abandoned fields with poor soil exhibit a dominance of *pioneering* herbaceous plants (the first plants to invade an area) such as chicory, dandelion, crabgrass, and common mullein. Large sections with mosses, lichens, and hard grasses may indicate severe nutritional deficiencies. Few mammals are present, the dominant animal forms being insects and birds.

On better soils, a more lush and varied plant growth develops. If surrounding conditions are of a natural, forested type, milkweed, clover, wild bergamot, butterfly weed, wild carrot, black-eyed susan, and chicory may be common. Scattered alfalfa and clover may indicate recent farming. Scattered pockets of sedges and rushes may indicate wet areas. The presence of grasses and forbs creates changes in the soil texture, structure, temperature, and moisture. Grasses may dominate the area for a while. Continued changes in the microclimate and the soil eventually allow the natural introduction of pioneer sun-loving trees and shrubs. Aspen and hawthorn trees along with gray dogwood and sumac shrubs become common sights scattered throughout the fields and along the old fencerows.

Songbirds and game birds such as sparrows, meadow larks, quail, and pheasant are common in this community. Large numbers of small mammals exist, though many are not evident because they are nocturnal or live among the grass roots. The cottontail rabbit and woodchuck are apt to be common. Typically in these early stages of plant-animal succession, the variety of species

FIGURE 7.5. Sun-loving trees and shrubs.

tends to be small, but the number of individuals in the species is usually large. (See Table 7.1, below, and Table 7.4, page 78.)

THE FOREST COMMUNITY

The sun-loving tree-and-shrub community, if undisturbed, is eventually replaced by a forest community of trees whose seedlings are shade-tolerant. This forest community represents a self-perpetuating last stage of plant succession unless some natural calamity such as a wildfire, sets it back to bare land or some earlier stage of succession. Depending on soil and climatic conditions, a site can have a variety of climax forest types. As a rule, forest communities can be characterized simply as having a dense canopy of treetops, a limited number of understory trees and shrubs beneath, and an abundant growth of herbaceous plants on the forest floor.

A variety of animal species occupy the forest, with woodpeckers, warblers vireos, squirrels, and raccoons among the most common. The mature forest can

Table 7.1. SOME PLANTS AND ANIMALS OF UPLAND FIELDS

ANNUALS AND PERENNIALS		SHRUBS	MAMMALS	BIRDS	REPTILES AND AMPHIBIANS
Goosefoot	Sunflower	Hazel	Ground squirrel	Pheasant	Garter snake
Smartweed	Thistle	Ninebark	Jumping mouse	Meadowlark	Fox snake
Pigweed	Ragweed	Hawthorn	Meadow mouse	Bobolink	Milk snake
Mustard	Quack grass	Blackberry	Least weasel	Field sparrow	American toad
Beggarstick	Bindweed	Wild rose	Cottontail	Vesper sparrow	Chorus frog
Aster	Chicory	Sumac	Skunk	Marsh hawk	Pheasant
Goldenrod	Yarrow	Willow	Mole	Northern yellow-throat	Red-bellied snake
		Dogwood			

Table 7.2. SOME PLANTS AND ANIMALS OF THE FOREST

PLANTS			ANIMALS		
TREES	SHRUBS	HERBS	MAMMALS	BIRDS	REPTILES AND AMPHIBIANS
Basswood	Prickly ash	Jack-in-the-pulpit	Woodland deer mouse	Sharp-shinned hawk	Wood turtle
Oak	Poison ivy	Wild geranium	Squirrel	Bluejay	Wood frog
Hickory	Dogwood	Trillium	Woodland jumping mouse	Warbler	Tree frog
Ash	Viburnum	Bloodroot	Weasel	Vireo	
Maple	Huckleberry	Wild ginger	Chipmunk	Thrush	
Elm		Mayapple		Woodpecker	
Ironwood		Violet			

serve the human community in a variety of ways, such as a park, playground, and noise buffer, in addition to supplying wood products and watershed protection. (See Tables 7.2 and 7.4.)

Lowland Communities

Lowland communities can be characterized as having soil that is saturated with water much of the year and may have high concentrations of organic soil material. These factors severely restrict development by humans, especially for construction purposes. The value of lowland communities cannot be overlooked, however, because they are very important in their ecological functions as wildlife habitats and as landforms that retain and purify water. Lowland communities can be classified simply as seasonally flooded lowlands, marshes, swamps, and bogs. In somes states and provinces, many of these communities can be

FIGURE 7.6. The forest community.

John Victory

John Victory

FIGURE 7.7. A cattail marsh and a dogwood swamp border the edge of an inland lake.

found close together, especially around inland lakes. In more arid regions, there are only a precious few.

SEASONALLY FLOODED LOWLANDS

Seasonally flooded lowlands can be found along the edges of streams, lakes, and oceans, and in poorly drained depressions on higher ground. The soil is generally well drained for most of the summer, but may be saturated during periods of heavy runoff. Vegetation will vary from large stands of flood-plain hardwoods to herbaceous growth.

MARSHES

Marshes are found along the shallow margins of lakes, ponds, and oceans, and in low, poorly drained land where water collects and settles for several months of the year. The surface soil is highly organic, and the vegetation consists of reeds, sedges, grasses, and cattails. Water depth varies from three inches to six feet, and in many locations there is open water.

SWAMPS

Swamps, like marshes, are usually found along the shallow margins of lakes and ponds and in low, poorly drained land where water stands for long periods of time. But, unlike the marsh, the swamp is characterized by woody vegetation, that is, shrubs and/or trees that can live in the wet, organic soils.

BOGS

Bog conditions are found where drainage out of an area is lacking and water is present all year, with water temperatures low for portions of the year. The low

Table 7.3. SOME PLANTS AND ANIMALS OF WETLANDS

PLANTS				ANIMALS		
FLOOD PLAINS	MARSHES	SWAMPS	BOGS	MARSHES	SWAMPS	BOGS
Maple	Cattail	Tamarack	Tamarack	Muskrat	Vole	Warbler
Ash	Rushes	White cedar	Leatherleaf	Mink	Shrew	Lemming
Basswood	Bur-reed	Spruce	Sphagnum moss	Ducks	Sparrow	Goose
	Pondweed	Maple	Sundew	Coot	Rail	Heron
Cottonsedge	Plantain	Alder	Pitcher plant	Redwing	Beaver	Otter
Oak	Waterlily	Willow	Bog kalmia	Deer	Raccoon	Rabbit
Sycamore	Arrowhead	Dogwood	Blueberry	Snake	Hawk	Owl
Hophornbeam	Grasses	Sumac	Grasses	Kingfisher	Wren	Grebe
		Birch	Sedges	Turtle	Frog	Toad

temperature and poor drainage create an environment where organic debris (dead plant material) only partially decompose, and they form peat. Resulting acid conditions limit the variety of animals and plants found in the bog. A major difference between a bog and a swamp is that the bog has wet, spongy ground; it shakes or yields under foot and is composed mostly of decayed and living mosses, principally sphagnum moss, often called peat moss. (See Table 7.3.)

EDGES OF PLANT-ANIMAL COMMUNITIES

Much development for which we must plan takes place in rural areas undergoing urbanization, where abandoned fields are often separated by woodlots and fencerows. A fencerow is characterized by a thin strip of vegetation that has grown up along the undisturbed edges of abandoned or cultivated fields. This edge may have been the property line of a farmer's land, initially defined by a plowed furrow, wire fence, or stone wall. A bordering area, public utility line, or a public right-of-way may also take on a similar appearance. As a general practice, the farmer does not turn over the soil right up to this line, but instead leaves a strip of land undisturbed. Over the years, this strip undergoes the normal steps in plant succession. Consequently, the strip may exhibit the characteristics of any one of a variety of plant communities or may include several such dryland and wetland types.

FIGURE 7.8. A typical fencerow.

John Victory

In proposing plans for changes in areas blessed with fencerows, woodlots, and other such remaining pieces of plant-animal communities, it is usually wise to plan for including them and safeguarding them during development. Once destroyed, they are often impossible to reproduce. They have great value for humans as well as for the other animals and plants with whom we share our planet.

A concept important in the discussion of communities is expressed by the term *edge*. Often the vegetative strip that divides field communities will reach to a forested area. There the plants blend with the edge of the forest. The fencerow and forest border both can be defined as edges, as can the edge of a marsh or any other plant-animal community. Therefore, edges, so important in wildlife management, can be defined as the places where two different vegetative or aquatic communities meet and more or less merge.

Theoretically, a greater variety of wildlife species and greater number of individuals of those species can be produced upon a given piece of land if edges are created. The creation of edges can be accomplished by such operations as blasting holes in marshes, cutting openings in forests, and planting sun-loving trees as a fencerow across a field.

Wildlife uses these fencerows as *travel lanes* when moving from one habitat or plant-animal community to another. Also, permanent residents of open-field communities may use a fencerow in traveling or as protective cover into which to escape. Many species, like the catbird, brown thrasher, vireo, and some

FIGURE 7.9. Forest edge.

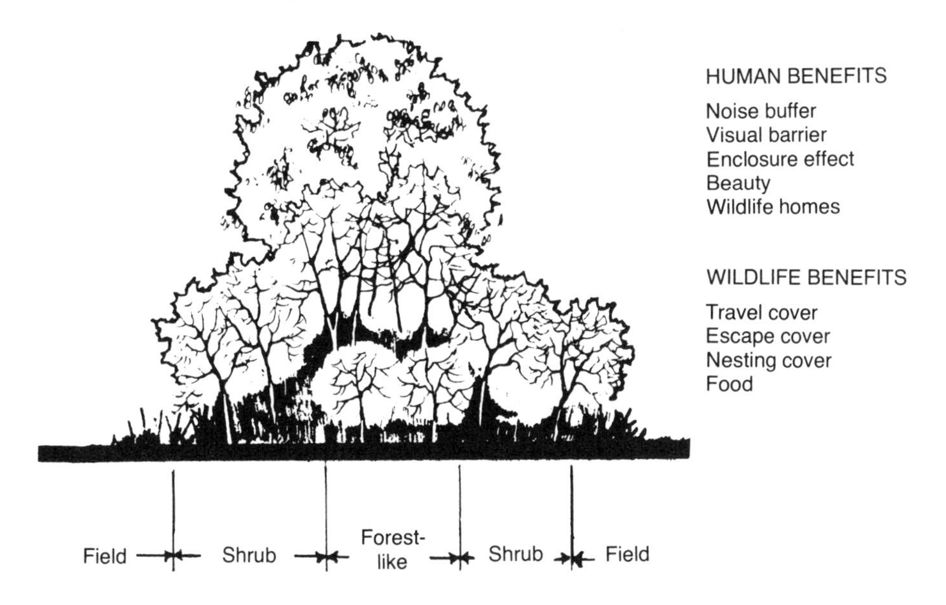

FIGURE 7.10. Benefits of a good fencerow shown in cross-section.

warblers also use such areas for nesting. You will notice the cross-section of the fencerow (Figure 7.10) that the types of plants that make up a fencerow vary from grasses and forbs to shrubs and trees. The variety of composition supplies a greater variety of food and cover for wildlife.

In northern and mountain states, aspen is a common tree of the fencerows. It is a major source of food for deer and grouse. Other plants, such as cherry (fruits), hickory (nuts), dogwood (buds and berries), witch hazel (seeds), sumac (fruits), and rose (fruits) are all excellent food for animals. (See chapter 12 for a list of many plants used as food by wildlife.)

Because fencerows aid animal survival in many ways, wildlife from surrounding habitats use the fencerow. Table 7.4, "Some Wildlife of Three Habitats," shows some of the animals that are found in field, fencerow, and forest habitats. This only begins to indicate the variety; an area with fencerows is attractive to many more species of animals than is an area without such edges.

In human-interest terms, a well-designed school site, shopping center, or subdivision should have adequate vegetation to provide for wildlife. Such vegetation can also serve to buffer noise, filter the air, and supply beauty spots and trails for passive recreation and education. Humans like plants. Plants make our environment a nicer place in which to live.

LANDSCAPE FUNCTIONS OF PLANTS

Plants are used extensively in our urban environment to beautify the external features of our homes and, to a lesser degree, our place of work. Grass, as lawn, is the most common plant used. Aside from being attractive, grass prevents erosion of soil, is clean, and is relatively easy to maintain. But there are many more ways in which plants can function to create a more livable environment.

Table 7.4. SOME WILDLIFE OF THREE HABITATS

MAMMALS		
UPLAND FIELDS	FORESTS	FOREST EDGES AND FENCEROWS
Woodchuck	Flying squirrel	Woodchuck
Cottontail	Red squirrel	Cottontail
Red fox	Gray squirrel	Red fox
Skunk	Fox squirrel	Skunk
Least weasel	Star-nosed mole	Least weasel
White-tailed deer	White-tailed deer	White-tailed deer
Short-tailed shrew	Short-tailed shrew	Short-tailed shrew
Jumping mouse	Jumping mouse	Jumping mouse
Prairie deer mouse	Opossum	Opossum
Meadow vole	White-footed mouse	White-footed mouse
House mouse	Woodland deer mouse	Woodland deer mouse
Ground squirrel	Raccoon	Raccoon

BIRDS		
UPLAND FIELDS	FORESTS	FOREST EDGES AND FENCEROWS
Pheasant	Warblers	Pheasant
Tree swallow	Vireos	Tree swallow
Quail	Brown creeper	Quail
Flicker	Tanager	Flicker
Mourning dove	Chickadee	Mourning dove
Crow	Crow	Crow
Bluejay	Bluejay	Bluejay
Red-tailed hawk	Red-tailed hawk	Red-tailed hawk
Upland plover	Common grackle	Common grackle
Field sparrow	Barred owl	Catbird
Horned lark	Grosbeak	Thrasher
Meadowlark	Woodpeckers	Cardinal
Savannah sparrow	Nuthatch	Yellow warbler
Bluebird		Song sparrow
Bobolink		Towhee
		Fox sparrow
		Cowbird
		Woodcock

Screens and Buffers in Your Landscaping Plans

Plants can function as screens to purify the air, eliminate objectionable views, screen out the sun, and cut down on the effect of noise. (See Figure 7.11.)

Microclimate Control

Closely related to the screening and buffering effects of plants is their ability to regulate and control the microclimates in our environment. Shade trees can maintain a cool climate in the summer. (See Figure 7.12.) Evergreen trees can keep buildings warmer in the winter by holding warm air on their sunny south side, providing shelter from winter winds.

Traffic Control in Developments

Plants can function as guides to help control vehicular or pedestrian traffic within a development. Plants can funnel people through a corridor or along a

FIGURE 7.11. Plants as visual screens and buffers against noise.

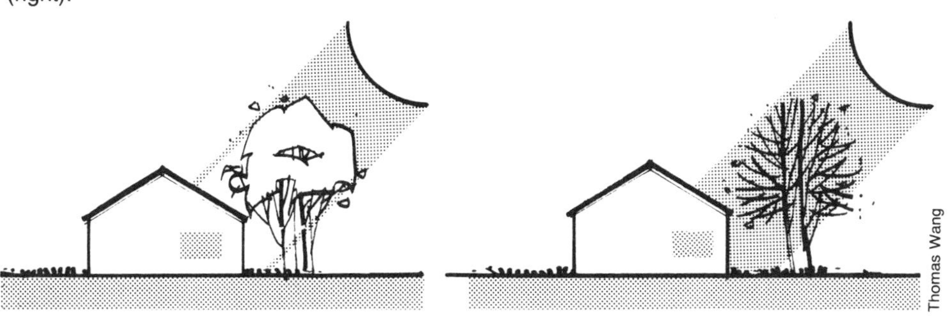

FIGURE 7.12. Plants used to control microclimates: cool in summer (left); warm in winter (right).

hedge toward a building or along a nature trail that leads to interesting places. Or they can block the flow of traffic that would create problems were it allowed to continue in a certain direction. (See Figure 7.13.)

Erosion Control, Often Neglected During Construction

Soil should be covered at all times to prevent its erosion by wind or water. This means grass, other ground-covering plant species, wood chips, gravel, or stone should be used as much of the time as possible during and after construction. Trees and shrubs, too, can help break the force of wind and rain, and help stabilize the soil with their roots. (See Figure 7.14.)

FIGURE 7.13. Plants as traffic control elements.

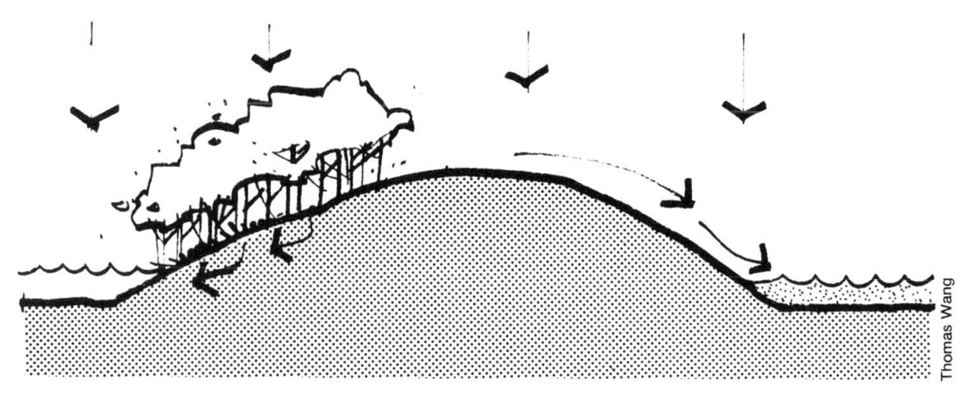

FIGURE 7.14. Plants used to control erosion: water cool and clear (left); water warm and dirty (right).

Spatial Relationships

The term *spatial relationship* refers to the effects that plants can have in creating spaces, much as walls do in a room. Humans perceive space primarily by sight. This perception is then internalized and has a direct effect upon how we feel. Think of how you would feel if you were in a small clothes closet. Then imagine the feeling you would have if you were in the middle of a football field all by yourself. The spaces would generate totally different feelings. Plants can be saved, maintained, or planted on a site to produce a variety of spaces. This is especially true for areas used by small children. (See Figure 7.15.)

Aesthetics

The aesthetic functions of a plant or cluster of plants lie in their ability to enhance the existing features of a development. Most often, these functions are directed toward complementing a building's architecture or toward unifying different, unrelated elements in a landscape. The use of plants' height, color, texture, or shape to make visual frames around objects of interest, such as scenes, statues, or fountains, is also a common practice, as is their use in framing vistas and panoramas of more distant landscapes. (See Figure 7.16.)

Not to be neglected in planning are the effects of plants on our nonvisual senses of smell (lilacs in bloom), touch (grass under bare feet), and sound (wind in pines).

FIGURE 7.15. Using plants to create interesting spaces.

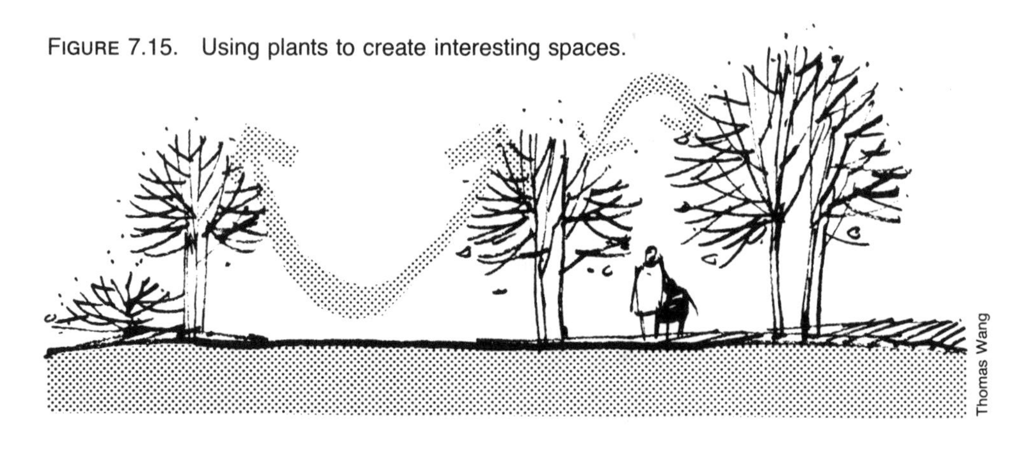

FIGURE 7.16. Plants can be used to unify elements in the landscape (left, arrows show connection) or to frame objects of interest (right).

INVENTORY AND ANALYSIS OF PLANT COMMUNITIES

Before any other site studies are conducted, a rough map of vegetation should be drafted. (See Figure 7.17.) Such a map should outline the size and shape of the different plant communities. This can be done by tracing their outline onto tracing paper or a base map at the site, in the laboratory, or at home. Aerial photographs can supply this information. If an aerial photograph is not available, the outline and location can be defined by using the survey traverse and triangulation methods of mapping discussed in Chapter 4.

Plant communities are usually determined and described by analyzing the kinds of species in the community, how many plants of the species are present (density), and the total space occupied by the species (coverage). In many cases, the plant species that are most visible from a distance give the name to the plant community. If the investigator is not familiar with plant communities, it may be

FIGURE 7.17. A rough vegetative map outlining the plant communities should be drawn before any field studies are carried out.

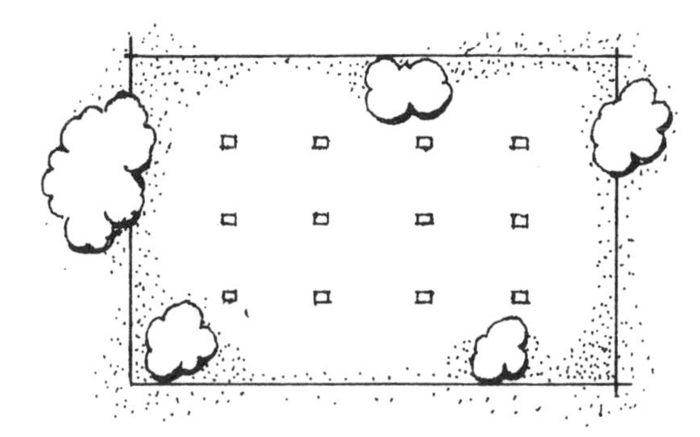

FIGURE 7.18. Open area with twelve quadrats staked out.

necessary to conduct a sample survey on the site to determine the dominant forms of plant life. Two types of surveys, quadrat and transect, are often used for these purposes.

Quadrat Sampling

Density is the number of organisms in a given area. Density can be determined by quadrat sampling. A *quadrat* is a plot of land that is assumed to be representative of the total plant community. Quadrats may vary in size, number, and arrangment in the site, depending on the nature of the study. One quadrat used in a study of a forest may be 50 × 50 meters; a study in a field may use quadrats one meter square. Quadrats can be systematically placed throughout the community in a gridlike pattern (see Figure 7.18) or scattered in a random fashion. In each case, the total number of plants is recorded by species. From these figures, density, relative density, and the total number of organisms for the entire area can be calculated. Once the study area has been established and the species counted, these factors can be arrived at by using the following formulas:

1. Density $= \dfrac{\text{Number of individuals of one species in area sampled}}{\text{Total square footage of area sampled}}$

2. Relative density $= \dfrac{\text{Density for one species}}{\text{Total density for all species}}$

3. Total number of organisms $=$ Number of individuals of one species $\times$ Total square footage of plant community

FIGURE 7.19. Three types of plants (A, B, and C) intercepting a transect line.

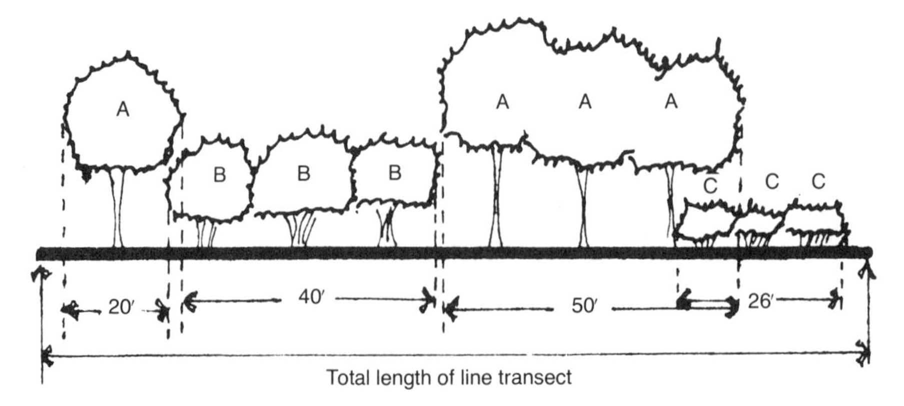

Line Transect Method (Line Intercept)

Another measurement that can be made is *plant coverage*, the percentage of area a species occupies in a community. The *line transect method* of sampling is often used to determine coverage. This method differs from the quadrat method in that straight lines are used instead of square plots. (See Figure 7.19.) Lengths of cord ten to fifty meters long are staked out across an area. The plants that touch a line are then identified, counted, and measured along the length of line that they occupy (that is, the plants intercept along the line). If a species is encountered many times along the transect, it may be considered a significant species in the community. Cover can be calculated by the following formula:

$$\text{Cover} = \frac{\text{Total intercept lengths for one species}}{\text{Total length of intercept}}$$

$$\text{Relative cover} = \frac{\text{Total of intercept lengths for all species}}{\text{Total of intercept lengths for one species}} \times 100$$

A slightly simpler method samples the *frequency* with which a species occurs in a community. The intercept distance of each plant is not measured, just counted.

Belt Transect Method

A *belt transect* is a long, narrow, rectangular plot that is divided into study blocks. It combines the quadrat and line transect methods of population survey. As in the previous methods, the number and length of the transect depends on the size of the area to be studied. This method is often used where the vegetation is varied and it is difficult to select a quadrat that typifies the area. As a rule, the total sample area in all methods should be about 10 percent of the area sampled.

The normal procedure in a belt transect study is to plot and stake out parallel lines (ten to thirty meters long) one meter apart across the boundaries of several communities. (See Figure 7.20.) The belt is then studied one section at a time. Site data are collected as in the quadrat and line transect studies, but the information can give a better picture of the general makeup of the different plant communities and the physical interrelationships that exist among them.

FIGURE 7.20. Belt transect.

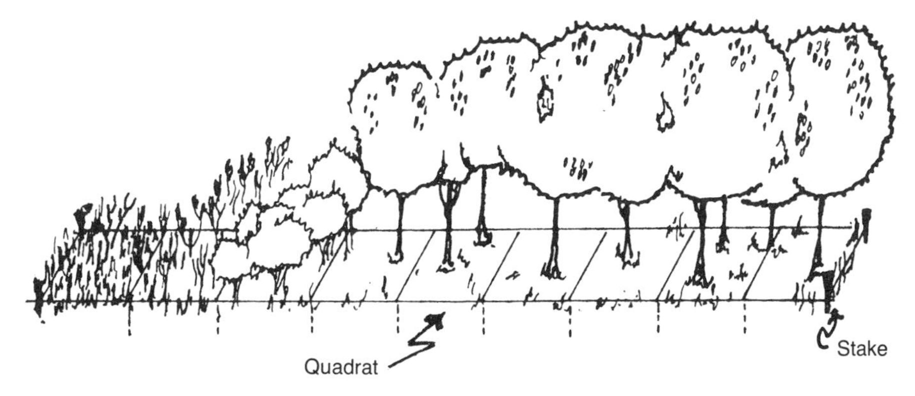

FIGURE 7.21

FORM FOR EVALUATION OF VEGETATION

Type of Development: Single-family homes

Major Development Features: 3,000 feet of road

50 homes

[and so forth]

Suitability Key

Excellent 1

Good 2

Fair 3

Poor 4

Plant community or Association	Water Retention	Aesthetics	Visual Screens	Noise Buffers	Spatial Benefits	Vehicular Control	Pedestrian Control	Erosion Control	Wildlife Habitats	Indicator of Soils	Rare-Fragile-Endangered

EVALUATING AND ILLUSTRATING VEGETATIVE INFORMATION

Site data collected during a vegetative inventory should be summarized and illustrated on a map of vegetation to serve as a quick and easy visual aid to express the kind, size, and location of the vegetative communities that exist on the site. Generally, an analysis of the data is also included when submitting a proposal for environmental changes. A *suitability evaluation form* can be helpful in analyzing and evaluating vegetation types for their functional qualities. Figure 7.21, "Form for Evaluation of Vegetation," is an example of a simple suitability form.

The final map is drawn from an aerial photograph and from the rough site map made during data collection. All the maps should have a key describing the graphic symbols. An analysis portion can show planners' value judgments about how the vegetative communities may respond to the changes caused by the proposed land changes. Any such statements can be included on the map if space permits or they can be made on a separate page.

Compare the following two vegetation studies. The map of Lathrup Woods (Figure 7.22) represents a study of a site conducted by a high school student. Included on his map is an inventory of conspicuous plants in their respective communities along with a brief text analyzing the vegetational history of the

FIGURE 7.22. Vegetation study illustrated by a student.

VEGETATIVE ANALYSIS

Lathrup Woods was a mature forest of beech-maple and oak-hickory prior to the arrival of humans. When the settlers came, the vegetation was altered by grazing, crops, exotic plants, woodlot activities, and disease. Periodic flooding also helped create a variety of plant species. As a result most of the area is in some stage of reverting to the climax forest type of a mixed hardwood stand. Following is a list of the successional stages:

A. FIELDS OF GRASSES AND FORBS
Goldenrod Plantain
Chicory Aster
Ragweed Clover

B. SHRUB AND GRASS COMMUNITY
Thornapple Bluegrass
Timothy Staghorn sumac
Rose Bindweed

C. LOWLAND FOREST
Willow Poison Ivy
Sycamore Button bush
Nettle Cottonwood

D. MATURE FOREST (mixed hardwood)
Oak Spice bush
Maple Prickly ash
Beech Ironwood

LATHRUP WOODS

NORTH

Site analysis conducted by: Paul Wilson 3/17/76
Environmental conservation class, spring 1976

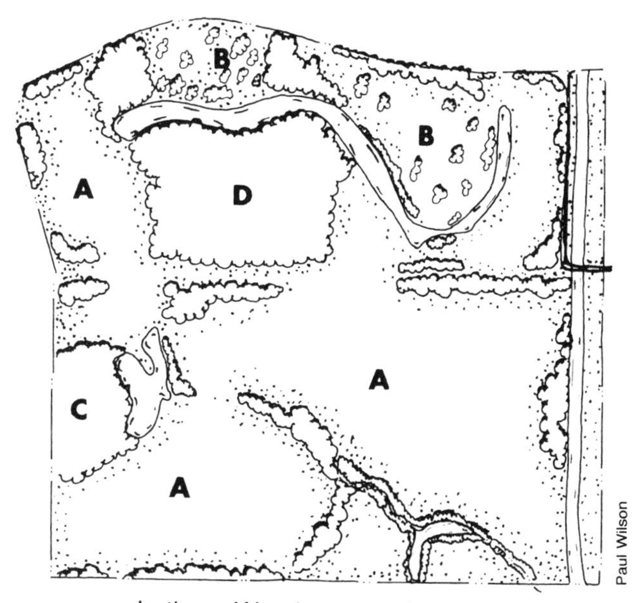

Lathrup Woods vegetative map.

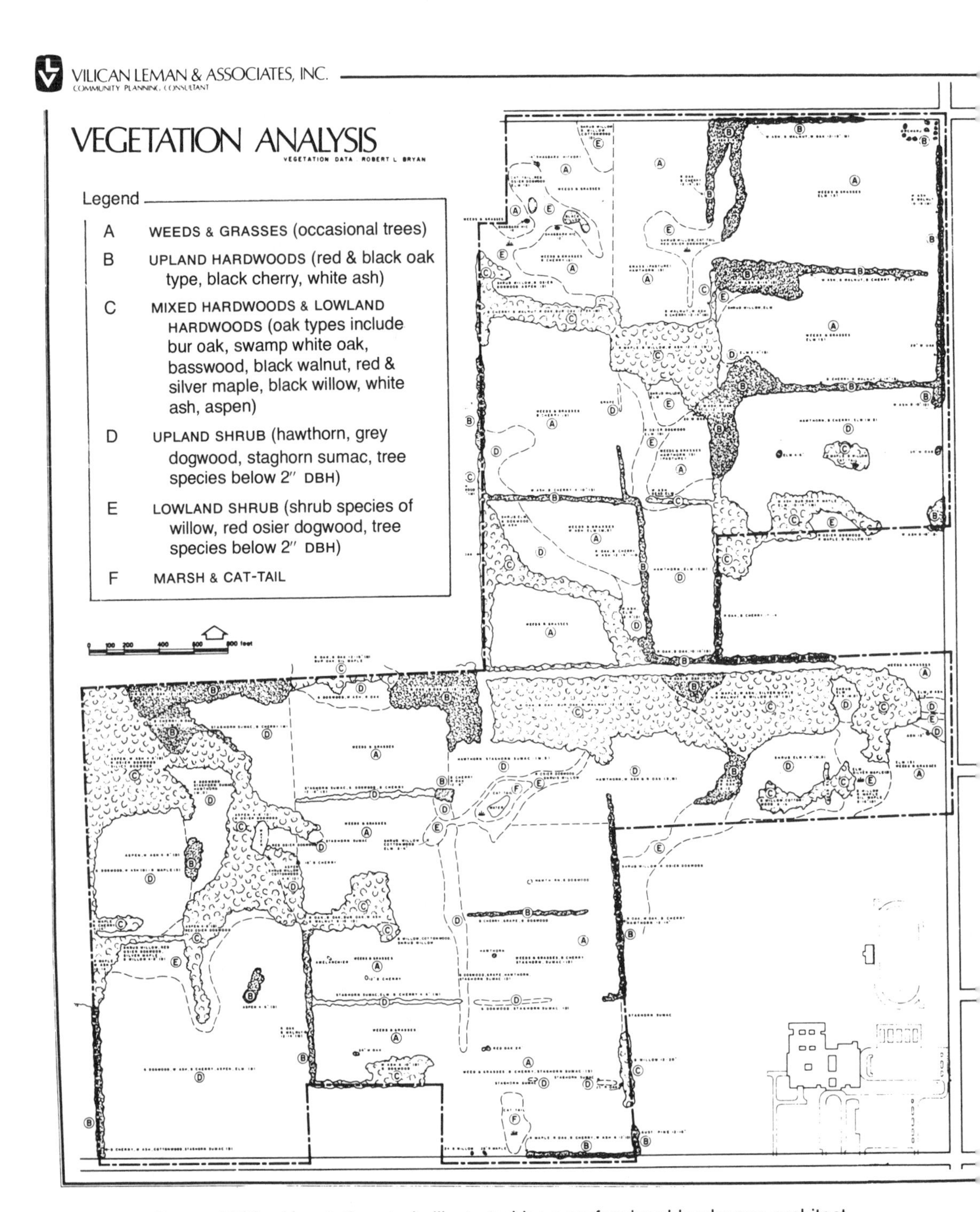

FIGURE 7.23. Vegetation study illustrated by a professional landscape architect.

site. Figure 7.23, a map also labeled "Vegetation Analysis," was done by a professional landscape architect. The vegetation inventory illustrated here was prepared as part of a comprehensive environmental study on a 377-acre site. The map was contained in a booklet summarizing the study. The analysis was included in the text of the booklet.

CONCLUSION

The study of plants and how they affect our environment can be interesting and rewarding. The popularity of horticulture and landscape architecture in our schools, as well as the many landscape services, flower shops, and greenhouses found in and around our urban centers, are testimony to the importance of plants in our environment. If there is a problem with our attitude toward plants, it probably results from our inability to appreciate the needs of plants. We must remember that plants, like humans and other animals, need space to grow along with adequate food and water. The following chapters can be helpful in understanding the complex relationships that exist between organisms and their environment.

REFERENCES

Bingham, Margorie T. *Flora of Oakland County*. Bulletin No. 22. Bloomfield Hills, Mich.: Cranbrook Institute of Science, 1945.

Reid, George K. *Pond Life: A Guide to Common Plants and Animals of North America*. New York: Golden Press, 1967.

Smith, Guy-Harold, ed. *Conservation of Natural Resources*. New York: John Wiley, 1966.

Smith, Robert L. *Ecology and Field Biology*. New York: Harper and Row, 1974.

Tobey, George B. *A History of Landscape Architecture*. New York: American Elsevier, 1973.

U.S. Department of the Interior. *Classification of Wetlands and Deepwater Habitats of the United States*. FWS/OBS-79/31. Washington, D.C.: 1979.

U.S. Department of the Interior, National Park Service. *Plants/People/and Environmental Quality*, by Gary O. Robinette. Washington, D.C.: 1972.

Watts, May. *Reading the Landscape*. New York: Macmillan, 1967.

Wisconsin Conservation Department. *Parade of Plants Series:* vol. 1, *Wetlands;* vol. 2, *Uplands Fields;* vol. 3, *Forest*, by Ruth L. Hine and George C. Knudsen. Madison, Wis.: 1963.

8 TOPOGRAPHY AND SOIL: INVENTORY AND ANALYSIS

The kinds of soils found on a site depend on their organic and mineral content. These factors determine the amount of air space and water space, as well as the amount of nutrients available for plants. All these factors—air space, water space, minerals, organic material, and organisms—make up the soil. Their interrelationship in the soil has a direct effect upon the kinds of plant and animal life present, the types of man-made structures that can be constructed, and other uses to which land can be put.

The topography of the land also plays an important role in the lives of plants and animals. Those that live on steep slopes encounter different conditions than those on flat ground. For example, the rays of the sun will be more direct during the year and will consequently supply more energy to soil and plants that are found on a south-facing slope. Because water drains off sloping land more rapidly, plants that live there must survive with less water.

Soil on the slope of hills is also subject to the erosive action of water. If the slope is steep, rain may wash away all the topsoil on the hill. This may continue until a subsoil is reached that perhaps withstands the erosion. Any structure proposed for such an area can be very expensive to construct and will add to the erosion problems, unless expensive measures are taken to alleviate the erosive action.

All these factors related to soil and topography must be considered when you propose a change on a piece of land. Information that will help determine what changes should occur can be expressed by conducting a soil inventory and illustrating the data with a soil map. Such an inventory may illustrate areas that are subject to flooding, or where the soil is well drained, poorly drained, or seasonally waterlogged. Your studies may suggest how susceptible the soil is to erosion, or point out the hazards of slippage, gullying, or shrinkage. They may also enable any engineers helping you to make good judgments on how deep the bedrock is and whether the soil is unstable and therefore susceptible to slippage or sinking that may resut in cracked foundations or caving.

For most projects, a soil analysis is done by a professional soil scientist. A thorough inventory generally consists of such studies as topography, slope, soil

FIGURE 8.1. Types of slopes: smooth and sloping (left); irregular and wavy (right).

depth and layering, soil permeability, texture, minerals, nutrients, and color. This information allows the soil scientist to classify the soil on a site into standard soil types, outline the extent of these soil types on a soils map, and evaluate the soil types relative to their capabilities for development.

TOPOGRAPHY: INVENTORY AND ANALYSIS

A soil scientist generally studies topography first. Topography is the configuration of the landscape, expressed as the type of slope and degree of slope. Both of these components are helpful in describing landscapes.

Types of Slopes

Type of slope refers to whether the land is smooth, flat, wavy, rolling, irregular, or highly irregular. (See Figure 8.1.) These elements can have an effect upon the way a land change is developed. Gently rolling land is desirable for a contractor because it is relatively easy to build on. Wavy, irregular land may be more attractive for homes because it offers more diversity and often better views, but it can be expensive to build on.

Percent of Slope

The slope, or lay, of the land affects not only the amount of erosion that may occur if the topsoil is disturbed but also the speed and amount of insoak and runoff, the ease of cultivation, and the cost of development. As the steepness of land increases, the uses of the land become more limited. Slope is expressed in percent, which represents the number of feet, meter, or other units of rise or fall in a hundred-unit horizontal distance. (See Figure 8.2.) The steepness of a slope

FIGURE 8.2. Percent of slope.

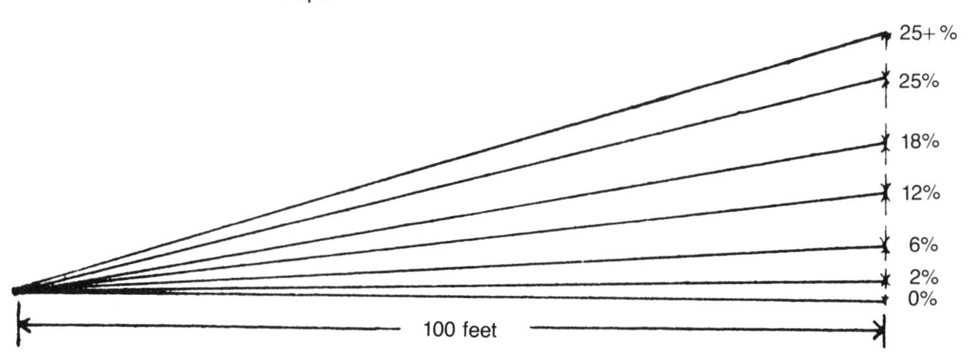

FIGURE 8.3. Students using a transit.

FIGURE 8.4. Measuring the percent of slope.

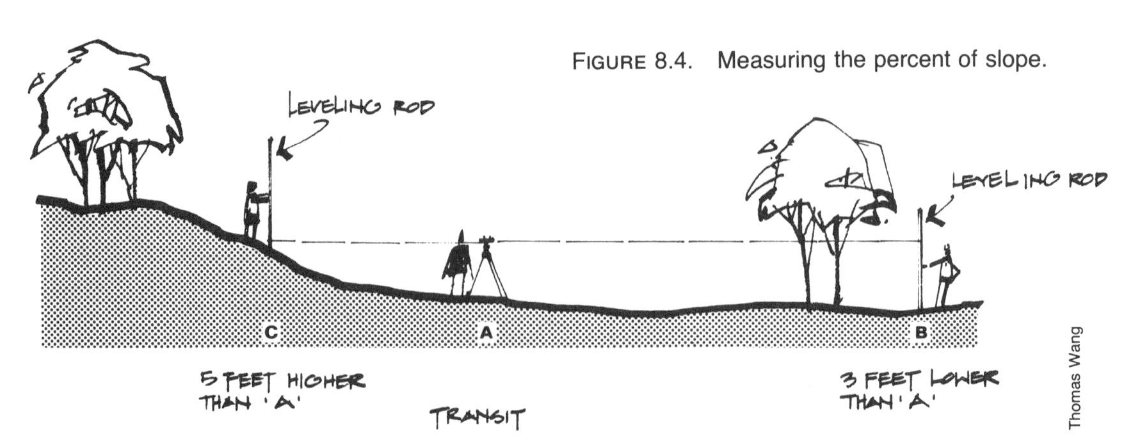

can be evaluated as follows, according to the United States Department of Agriculture's Soil Conservation Service:

Nearly Level (0–2%). Has no limitation on its uses. Any limitations are the result of other factors, such as drainage.
Gently Sloping (2%–6%). Desirable for almost any type of development; may have erosion problems; limitations are due mostly to factors other than slope.
Moderately Sloping (6%–12%). May have severe erosion problems and has a strong appeal for single-family development.
Strongly Sloping (12%–18%). Has severe limitations for all types of construction. Residential development is sometimes considered because of the scenic views associated with such terrain, or when other sites are unavailable.
Steep Slopes (18% and over). Undesirable for most development. May be suitable for split-level and earth-sheltered buildings, with expensive precautions.

Measuring the Percent of Slope

The percent of slope can be accurately measured with the use of a surveyor's transit, alidade, Abney level, or homemade sighting instruments made from a carpenter's bubble level.

A transit is the most accurate instrument for measuring slope. A transit is a telescopic sighting instrument mounted on a tripod that has adjustable legs and gears for leveling the telescope. Once leveled, the telescope can be turned in a circle and focused on the stadia rod, a giant ruler marked off in feet and inches or other units. The surveyor's helper holds it stationary anywhere within the sighting range of the transit, a spot usually 100 to 150 feet away, and relative height is then determined by the surveyor sighting on the rod.

The surveyor's transit gives the height of a spot on the landscape relative to where the transit is stationed. For example, Figure 8.4, "Measuring the percent of slope," shows that Point C is five feet higher than Point A, and Point B is three feet lower than Point A. If Point B were one hundred feet from Point A, the slope would be 3 percent.

Each spot on the landscape sighted is known as a spot elevation. Usually a large number of spot elevations are determined for each site. The elevations are plotted on a map from which the percent of slope can be calculated or contour lines drawn between spots to illustrate the contours of the landscape.

SOILS: INVENTORY AND ANALYSIS

After an inventory of the topography is made, a survey of the site must be conducted to identify the various types of soil, their location, and their extent. To carry out these studies, you must first become familiar with the basic characteristics and ecology of soils.

Characteristics of Soils

The most common element in soils is rock material. It can range from boulder- and pebble-sized particles to gravel, sand, silt, and clay. The most important

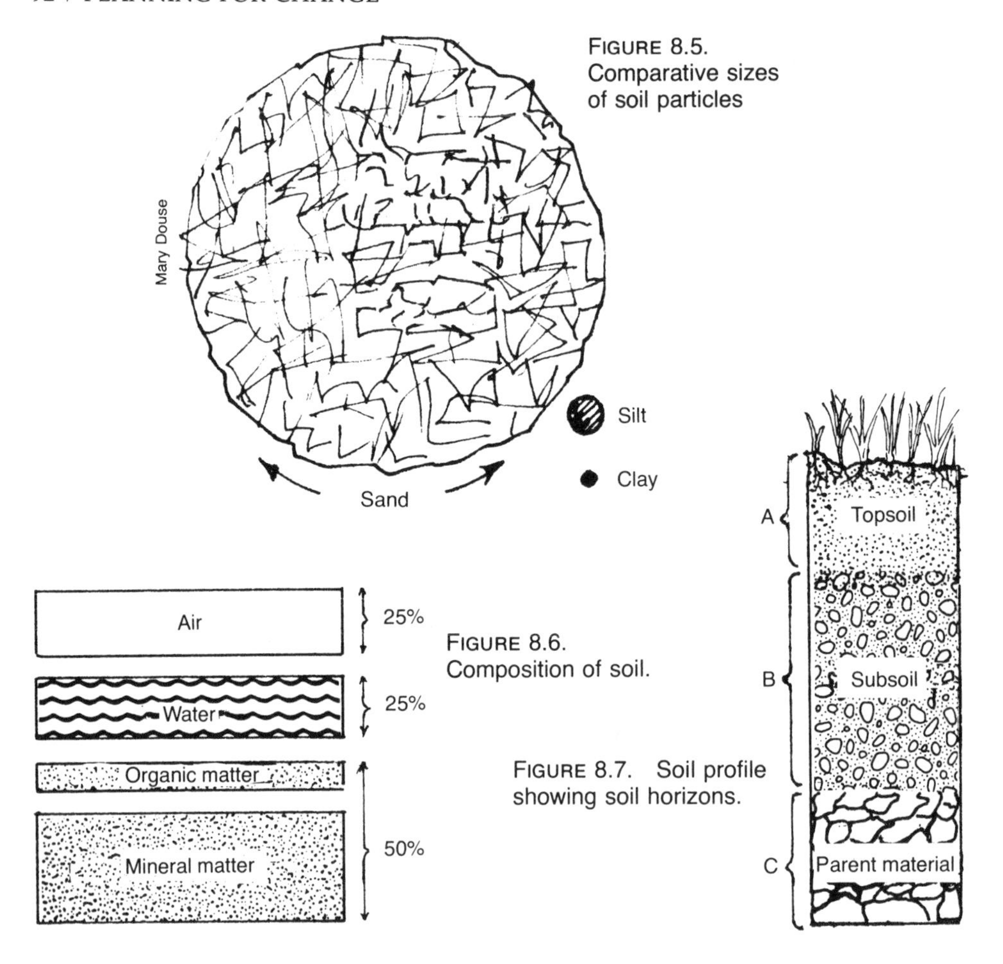

FIGURE 8.5. Comparative sizes of soil particles

FIGURE 8.6. Composition of soil.

FIGURE 8.7. Soil profile showing soil horizons.

sizes from the standpoint of soil classification are sand, silt, and clay. Figure 8.5 illustrates the approximate size of these particles. A sand particle can be as much as 625 times as large as a clay particle. Usually soils have different combinations of the three sizes. This combination is referred to as the soil's *texture*.

The amount of space for water and air in a soil largely depends upon the combination of these particles. A good soil capable of growing a wide variety of human-interest plants should have approximately 25 percent air space, 25 percent water space, and 50 percent mineral and organic matter. (See Figure 8.6.) Sand allows for adequate air space for plant and animal respiration. The clay holds sufficient moisture for living processes. Dead and decaying plant and animal matter (*humus*) add nutrients to the soil as well as water-holding qualities. Most of the humus is deposited in the top layer. With this addition of humus to the rock material, the soil is considered true *topsoil*.

Soil is characterized by the development of recognized layers called *horizons*. The top layer (*topsoil*) is the A Horizon. The B Horizon (*subsoil*) is a zone of clay accumulation that has washed down from the topsoil. The C Horizon is known as *parent material*; A and B Horizons were developed from it. (See Figure 8.7.) In conducting soil studies, it is important to understand the characteristics of all these layers, because the stability of any man-made structure depends upon a firm foundation.

Ecology of Soil

The composition of soil particles, the amount of organic matter, and the depth of the various horizons can vary from place to place, even in a relatively small area. The underlying rock material often determines the kind of rock particles in the soil. In many parts of North America, the soil is glacial material called *till*, left behind as the glaciers retreated northward. Farther south, *residual soils* developed from underlying rock unmodified by glaciation. In the Midwest, large areas have wind-deposited soil blown out from arid regions farther west.

Through the centuries, plants added humus to the rock material. Physical, chemical, and biological activities reduced the larger rock particles to smaller and smaller sizes. Erosion carried the finer particles to lower levels in the landscape. Low, undrained wet areas allowed decaying plant material to build up, causing thick deposits of dark peat with few rock particles evident. Higher, dryer ground sometimes lost its smaller rock particles through erosion, leaving behind a sandy soil. Given enough time and humidity for plant life, a dark layer of topsoil could develop.

Because of the porosity of the soil, water will readily pass through topsoil. As the water percolates down, it leaches out soluble minerals and fine clay particles and deposits them in the soil below. The color of each horizon of soil depends upon the minerals in it, the amount of organic matter, and the relative amount of air and water in the pores between the mineral particles. Soil color can tell you a lot about the soil's relation to its environment.

Outlining Soil Deposits

A fairly accurate description of the size and configuration of a soil deposit can be obtained by using a vegetation map or aerial photograph, and a series of checks on the site. Plant communities almost always directly reflect the soil and water conditions upon which they grow. Some particularly helpful species are referred to by ecologists as *plant indicators*; examples are red maple, indicating moist soils in the East, and tamarisk (salt cedar), outlining alkaline wet areas in the Southwest. In many ways the relationships between soil and water that guide the growth of plants also have a direct impact upon man-made structures such as buildings and highways. It may be helpful to review the major vegetative types and their position on the site, because some of the major soil types correspond to these communities.

After obtaining a vegetative or aerial map for your site, outline the major plant communities on it. Then plant specimens for identification should be collected within these rough outlines. Soil samples should then be taken to determine the composition of the soil in these areas. Your final soil map will thereby be based on your vegetation map, refined by soil sampling on your site.

Careful sampling of the soils at appropriate spots will allow you to evaluate the soils adequately. Sampling must be carried out at different depths. Consequently, a soil auger, a shovel, or a post-hole digger is required to reach down into the soil layers. The location of the collecting stations must be recorded on the soils map, and the data collected at the station should be recorded on a

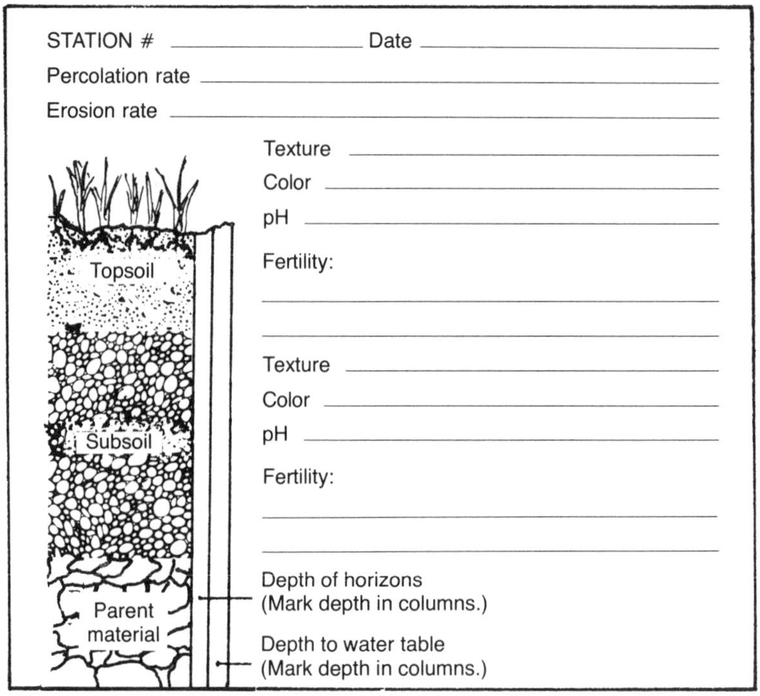

STATION # ______________ Date ______________

Percolation rate ______________

Erosion rate ______________

Topsoil

Texture ______________

Color ______________

pH ______________

Fertility:

____

____

Subsoil

Texture ______________

Color ______________

pH ______________

Fertility:

____

____

Parent material

Depth of horizons
(Mark depth in columns.)

Depth to water table
(Mark depth in columns.)

FIGURE 8.8. Chart of soil profile at a sampling station.

chart. A *soil profile chart* can be constructed for each station. (See Figure 8.8.)

DEPTH

Some of the water that falls as rain or snow travels down through the soil until it reaches a layer of rock or soil it cannot penetrate. Water accumulates above this impervious layer. The top of the saturated layer of soil is called the water table. *Depth* measures the distance below the surface at which water saturates the soil. The water table may be near the surface or very deep, depending on such environmental factors as rainfall, evaporation, transpiration by plants, and pumping by people. It also may vary markedly with the seasons, so water table depth should be sampled at both a wet and a dry season.

The water table can have considerable effect on the use of a site. *Shallow soils* (with a high water table or high impervious layer of soil) may make the site unsuited for buildings or septic systems. *Deep soils* (with a low water table or low impervious layer of soil) can be ideal for many purposes. Following is a rough guide for evaluating depth.

Deep Soils (40 inches and over). Usually excellent for water storage and plant growth; good for most types of development.

Moderately Deep Soils (20–40 inches deep). Considered good for storing water, drainage, and growing common plants.

Shallow Soils (0–20 inches deep). Generally store water and drain poorly and are too wet for many common plants; poor for development.

If a soil permits, the quickest method for determining depth is with a long, thin rod and a post-hole digger. The rod can be driven into the ground by hand to find out if an impervious soil layer exists near the surface. If none is present, a post-hole digger can be used to extract soil to a depth of 40 inches. If water is present, it will seep into the hole. The depth to the top of the water table can then be measured.

EROSION

Erosion is the process of losing soil. It occurs as a result of wind and/or water action on the surface of the soil. It can be accelerated by the loss of protective plant cover, nutrients, and humus from the topsoil. *Sheet erosion* is the gradual loss of soil by water action, with little or no visual evidence of soil being lost. Erosion becomes more evident when rills and gullies are formed. The degree of erosion can be classified as follows (per the U.S. Soil Conservation Service):

Slight Erosion. All or nearly all the original surface soil may still be present.
Moderate Erosion. A mixture of original surface soil and subsoil is present.
Severe Erosion. Most topsoil removed, leaving soil visible. Rills, gullies, or wind blowouts may be present.
Very Severe Erosion. Severe gullies or wind blowouts are present.

Slight to moderate erosion may be difficult for an untrained person to detect by casual observation. One way to obtain data on the amount of erosion that

FIGURE 8.9. Example of gully erosion in its beginning stages.

John Victory

has occurred is to compare profiles of the soil collected with a soil auger at various spots on the site. When site studies are concluded, it is simple to compare the various soil profiles collected throughout the site in order to determine the degree of topsoil lost in eroded areas.

Severe erosion is more easily detected. Rill erosion is caused when runoff is heavy and water concentrates in rivulets. Small grooves in the soil form and eventually reach several inches in depth and width. Generally, a rill becomes a gully when it exceeds 4 inches in depth.

TEXTURE

Texture refers to the percentage of sand, silt, and clay particles in the soil. Texture is important because soils are classified and named for their textural characteristics. Texture also determines the amount of water and air available for plant and animal life in the soil.

Like all organisms (except some specialized bacteria), plants and animals living in the soil need air and water. The spaces between the soil particles are normally occupied by air. The water used by soil organisms is usually found as a film on the soil particles themselves. When it rains and water passes through the soil, a thin film of water adheres to the soil particles. This water is the major source of water for soil organisms.

In sandy soil the pores between grains are large, resulting in the evaporation of water on the soil particles. Small clay particles, on the other hand, offer more surface for water, with a reduction in the size of the air spaces. As a result, a soil with fine particles ("fines") has a greater capacity for holding water than does a coarse soil. However, in clay soils a "cementing" reaction forms a compact crust, preventing water and air from entering. As a result, a certain ratio of sand and clay is required to achieve a proper balance of water and air for most plant and animal life.

Soil classes are determined by the percentage of sand, silt, and clay a soil contains. A *textural triangle* (like Figure 8.10) can be used to identify a soil class. A class of soil is determined by locating the percentage of sand, silt, and clay along the sides of the triangle and extending these lines into the triangle until they cross. A soil composed of 30 percent clay, 30 percent silt, and 40 percent sand would be classified as a clay loam. A soil composed of 10 percent clay, 70 percent silt, and 20 percent sand would be a silt loam.

Further description is often made by grouping these classes into larger, more general groups. Many soil maps follow this classification (per the U.S. Soil Conservation Service):

Fine-Textured Soils. Clays, clay loams, silt clay loams, and sandy clay loams. Made up mostly of clay and silt, they have a smooth feel and are sticky when moist. They can also be made into a ribbon when a small lump is pressed between the thumb and forefinger.

Medium-Textured Soils. Silt loams, loams, and fine sandy loams. Made up mostly of silt and clay or very fine sands, the soil may be gritty to the touch but still have enough clay and silt to make it hold together in a mold when moist.

Moderately Coarse-Textured Soils. Sandy loams and loamy sands. Made up

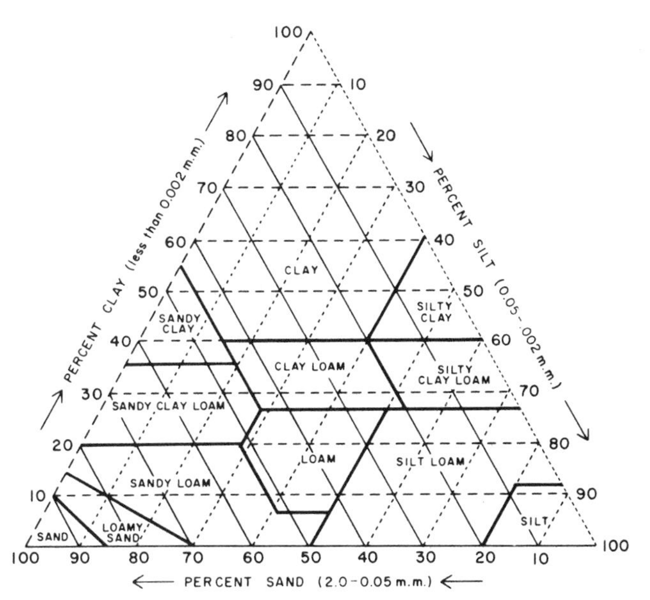

FIGURE 8.10. Soil triangle.

mostly of different-sized sand particles, these soils are difficult to mold when wet.

Very Coarse-Textured Soils. All sand or sand with gravel. They do not mold even when moist and feel gritty when rubbed between the fingers.

Organic Soils. So-called muck and peat soils. These are mostly composed of fibrous and/or woody matter, which is the result of the ground being saturated with water for most of the year.

It is relatively simple to determine the texture of a soil. A generally accepted procedure can be carried out on the site by taking a small portion of moistened soil and rubbing it between the fingers. The clump of soil should react in one of the above-mentioned ways. Sandy soil is difficult to mold and gritty to the touch; clay soils are easy to mold and slippery and plastic when moist; and silt soil is easy to mold but not so slippery or sticky as clay.

A more accurate determination can be made by observing what happens to a soil sample when it is shaken vigorously in a jar with water and allowed to settle out of the water. Different-sized particles will settle to the bottom at different rates because of their weight. Generally, the different-sized particles are also different in color, so that colored layers become evident.

COLOR

Minerals, moisture, oxidation, weathering, and the amount of organic content in the soil can all affect its color. Color is a good, quick, on-site indicator of texture and water relationships, but it is difficult to generalize about soil color because a color that indicates a high organic content and poor drainage for a northern Podzolic soil may suggest something different for Chernozems and Prairie-earths. Consequently, you should be familiar with the relationships between color and soil for your own area. Most counties throughout the United

States have a soil conservation district office of the U.S. Department of Agriculture where such information can be obtained.

PERMEABILITY

Permeability determines the rate at which water enters and moves through the soil. A knowledge of permeability can be useful in many ways. To build a home with a septic system, a permit from the town or county public health department is necessary. Officials will conduct percolation tests on the site before issuing a permit. If water does not drain (percolate) through the soil adequately, the landowner may not be allowed to build unless permeable soils are trucked in at considerable expense.

This procedure is necessary to insure that the ground water is not contaminated by human waste from the house. A septic system is a miniature sewage-disposal plant. Sewage from the home is broken down by bacteria in a septic tank. The waste water then drains slowly from the tank into a drainage field. The drainage field is composed of tile pipes with holes on the bottom, set upon a bed of gravel. Holes in the drain tile allow the waste water to seep slowly into the soil and eventually into the ground water. Soil organisms help purify the waste water before it reaches the water table. (See Figure 8.11.)

A quick site test for permeability can be carried out by placing in the ground a large can (for instance a No. 10 cylindrical) with both ends removed, filling it with water, and recording the time it takes for the water to percolate down into the soil. The percolation rate can then be used to evaluate the soil for possible building problems. The following rate categorization is from the U.S. Soil Conservation Service.

Very Slow (0–½ inch per hour). Indicates a high water-table, an impervious layer of subsoil, or finely textured soils. All would have limited building potential if a septic system were to be used.

Slow (½–¾ inch per hour). Might suggest much the same as very slow percolation rates.

Moderate (¾–2½ inches per hour). Suggest a moderately textured soil generally good for a septic system.

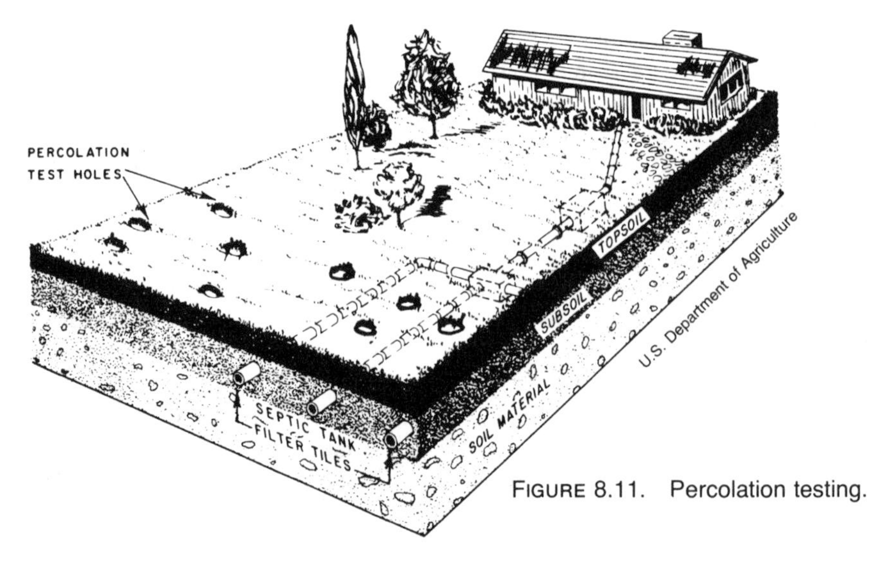

FIGURE 8.11. Percolation testing.

Rapid (2½–5 inches per hour). Indicates a moderately coarse-textured soil good for a septic system.

Very Rapid (5 inches or more per hour). Indicates a very sandy and gravelly soil good for a septic system. However, note that a porous soil may let water flow underground to contaminate an aquafer of potable (drinkable) water.

FERTILITY

The purpose of testing soils for fertility is to identify the major types of nutrients present (nitrogen, phosphorous, and potash) and the soils' acidic or alkaline qualities. This information can then be applied to planning future plantings. Because landscape planting is secondary in most development plans, this study is seldom done on large projects. On small projects, such as school sites or planning for homes, soil fertility tests can be important.

Soil is made up of many different materials, but for the purpose of understanding how soil nutrients and soil acidity are related, let us assume that soil is composed of millions of very small soil particles plus ions in solution. Ions are the smallest form of soluble nutrients used by plants. The surfaces of these particles are electrically charged, some negatively and some positively. A negatively charged particle attracts nutrient ions that have a positive charge. Negatively charged nutrient ions are attracted to positive soil particles. Once a nutrient ion is attracted to a soil particle, it is said to be adsorbed by the soil particle and stays attached until a stronger force removes it (for example, diffusion into root hairs of a plant).

When an ion is removed from a soil particle, the vacated spot on the soil particle may be occupied by another nutrient, or possibly a hydrogen ion. If a hydrogen ion is adsorbed on the soil particle the soil will be acid. Many nutrient ions, however, are basic ions, and therefore if a basic nutrient ion is adsorbed by the soil particle the soil will be alkaline.

If a soil is farmed for a long time, certain nutrient ions will be adsorbed by the plants. If nature cannot replace this supply, their position on the soil particles will be occupied by ions unusable by the plants that people want to grow. A fertile soil will supply the complete dietary needs of a growing plant. Each nutrient performs a specific function for the plant, and if these nutrients are absent, a deficiency will occur. Consequently, a chemical analysis of soils should be conducted to determine what nutrients are available or what should be added.

A soil may have sufficient nutrients, but they may not be available to the plants because of their acid or alkaline environment. This chemical environment is known as soil pH and is measured on a scale of 0 to 14 with 7.0 being neutral. A soil with a pH reading below 7 is acidic. A soil with a pH reading above 7 is basic.

Soil fertility, then, refers to the relative abundance of soluble organic and inorganic compounds in the soil that are available for plant growth, as well as the pH of the soil, which affects the ability of the plants to absorb nutrients.

Simple and inexpensive soil-test kits can be purchased to help you determine the fertility of the soil relative to the pH and such soil nutrients as nitrogen, phosphorous, and potash. Whatever soil kit you use, collect small soil samples in separate jars or bags for testing in the laboratory.

FIGURE 8.12

FORM FOR RECORDING SOIL DATA

SOIL TYPES

Surface		Subsoil
	FINE	
☐	Clay, clay loam	☐
☐	Silty clay loam	☐
☐	Sand clay loam	☐
	MEDIUM	
☐	Silt loam, loam	☐
	MODERATELY COARSE	
☐	Sandy loam	☐
☐	Loamy sand	☐
	COARSE	
☐	Sand	☐
	ORGANIC	
☐	Mucks and peats	☐

TYPE OF LANDSCAPE

☐ REGULAR
Uniform, simple, smooth

☐ IRREGULAR
Uneven, complex, wavy

SLOPE STEEPNESS

☐ Nearly level
0–2 ft. fall in 100 ft.

☐ Gently sloping
2–6 ft. fall in 100 ft.

☐ Moderate slope
6–12 ft. fall in 100 ft.

☐ Strongly sloping
12–18 ft. fall in 100 ft.

☐ Steep
18–25 ft. fall in 100 ft.

EROSION BASED ON PRESENT SURFACE SOIL

☐ SLIGHT
Mainly original surface soil

☐ MODERATE
Mixture of original top and subsoil

☐ SEVERE
Mainly subsoil, may have gullies

☐ VERY SEVERE
Severe gullies of deep blowouts

SOIL DEPTH TO GROUND WATER OR IMPERVIOUS LAYER

☐ Deep soils
Over 40 inches deep

☐ Moderately deep soils
30–40 inches deep

☐ Shallow soils
Less than 20 inches deep

CONDITION AND COLOR OF TOPSOIL

CONDITION	DARK (Gray, brown black)	MODERATELY DARK (dark brown to yellow brown)	LIGHT (pale brown-yellow)
Amount of organic material	☐ Excellent	☐ Good	☐ Low
Erosion factor	☐ Low	☐ Medium	☐ High
Aeration	☐ Excellent	☐ Good	☐ Low
Available nitrogen	☐ Excellent	☐ Good	☐ Low
Fertility	☐ Excellent	☐ Good	☐ Low

CONDITION AND COLOR OF SUBSURFACE SOIL

CONDITION	COLOR
Waterlogged soils and poor aeration	☐ Dull gray if in low rainfall areas
Well-drained soils	☐ Yellow, red-brown, black if in forest soils
Somewhat poorly to poorly drained soils	☐ Mottled gray if in humid soils

FORM FOR RECORDING SOIL STUDIES

While you are working at a site or in a laboratory, it can be helpful to have a prepared form for recording data. (See Figure 8.12.) It summarizes the studies that should be carried out at each station. It also contains brief statements to assist in analyzing the soil samples. The guide can be especially helpful to the beginner, who may find the many soil factors confusing at first.

ILLUSTRATING AND EVALUATING SOILS

From notes made at your site and from rough soil maps, a finished soils map should be prepared. This map should analyze the soil deposits that were found on the site, including their types. The soil map of Lathrup Woods (Figure 8.13) has an analysis included on the map. The map was prepared by a student in a high school class. Figure 8.14 is a map with a brief legend. The map was prepared by professional planners and was included in a comprehensive study of a 377-acre site. Statements analyzing the soil's suitability for construction were included in an accompanying booklet.

The degree and nature of each analysis varies with the project and the experience of the investigator. A site projected to support a large building or homes with septic systems will require a thorough analysis, whereas one destined to become a natural area may demand a less exhaustive study, unless nature studies are to be promoted there.

FIGURE 8.13. Soils map illustrated by a high school student.

SOIL ANALYSIS

The parent material for Lathrup Woods is glacial till, a mixture of sand, silt, and clay. Past cultivation and pasturing along with extraction practices have depleted the topsoil of its nutrient quality. Lying fallow over the years, plant material and animal activities have added a small amount of humus.

A. The medium-fine texture of these soils, in combination with low relief and high water table, severely limits intensive development. If not covered, vegetation could be destroyed by compaction.

B. Silt loam soils, forest covered with heavy humus layer.

C. Man-made land of rubble, sand, silt, and clay. Deposit is relatively deep, allowing for development of any type.

D. Organic soils of the marsh area. Water high, often exposed. Development depends upon the removal of soil, but depth of deposit suggests removal would be expensive.

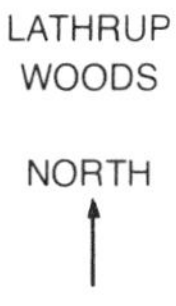

Site analysis conducted by: Paul Wilson 3/17/76
Environmental conservation class, spring 1976

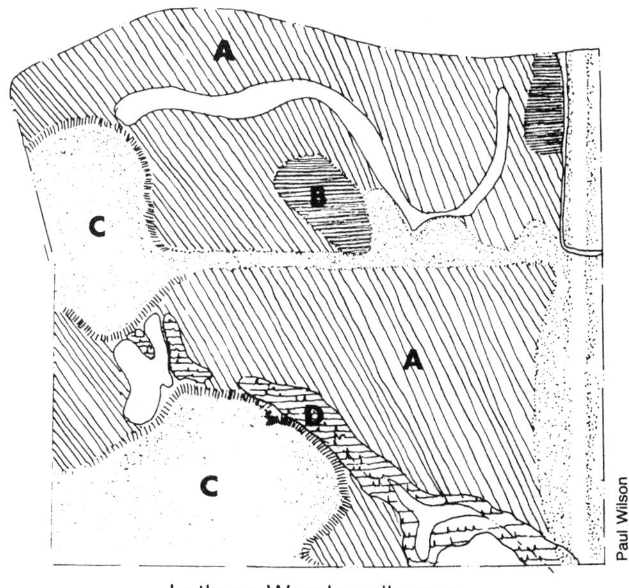

Lathrup Woods soils map.

The characteristics that affect soil's suitability for development are fairly well known. These characteristics can be compared with a developer's requirements and then a judgment can be made. Table 8.1, "Soil Suitability Guide" can serve as a reference for these comparative judgments.

FIGURE 8.14. Soils map illustrated by a professional landscape architect.

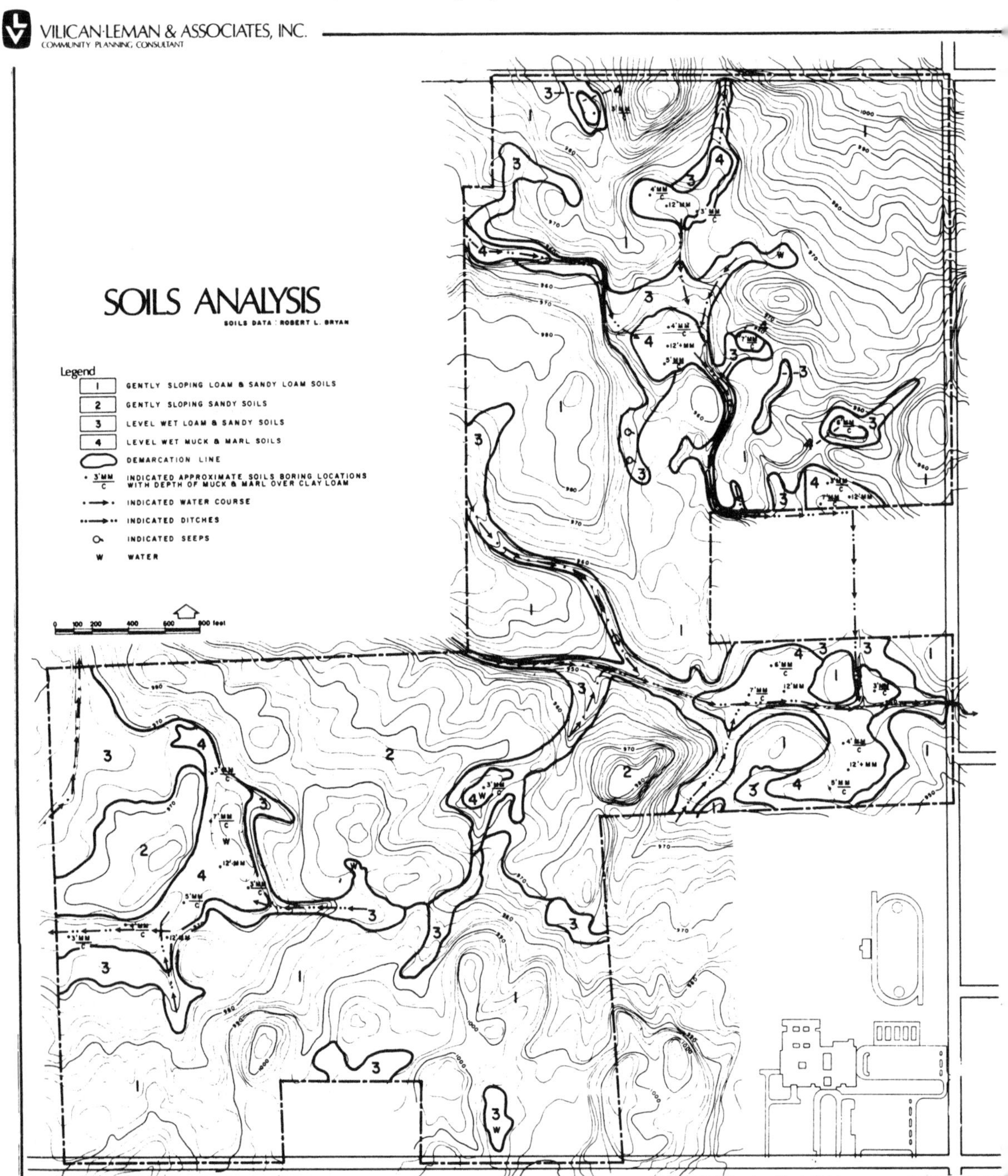

Table 8.1. SOIL SUITABILITY GUIDE

	AGRICULTURE AND CROPS	NATURAL RESOURCES—OPEN SPACES AND WILDLIFE	URBAN RECREATION FACILITIES	RESIDENTIAL—HOMES WITH SEPTIC SYSTEMS	INDUSTRIAL–COMMERCIAL AND HOMES WITH SEPTIC SYSTEMS	ROADS AND STREETS
1. Slope						
Nearly level	I	I	I	I	I	I
Gentle slope	II	I	II	II	II	I
Moderate slope	III	I	II	III	III	II–III
Strong slope	U	I	U	III	U	U
Steep slope	U	I	U	U	U	U
2. Texture: Topsoil						
Fine	I	I	I	III	II	II
Medium	I	I	I	II	I	I
Moderately Coarse	II	I	II	I	I–II	I–II
Coarse	III	I	I	I	II–III	II–III
Organic	III	I	U	U	U	U
3. Color: Topsoil						
Dull	II	I	II–III	III–U	III	U
Mottled	I	I	II–I	II–III	II	II–III
Bright	III	I	I	I	I	I
4. Depth						
Deep	I	I	I	I	I	I
Moderately deep	II	I	I	I	I	I
Shallow	III	I	II–III	III	III	U
Very Shallow	U	I	III–U	U	U	U
5. Erosion						
Slight	I	I	I	I	I	I
Moderate	II	I	II	II	II	I
Severe	III	I	III	III	III	III
Very Severe	U	I	U	U	U	U
6. Permeability						
Very slow	II	I	II–III	U	III	III–U
Slow	I	I	I–II	II–III	II	III–U
Moderate	I	I	I	I	I	II
Rapid	II	I	III	I	I–II	I
Very rapid	III	I	U	II	II–III	I

Source: Adapted from Robert W. George, Arnold Mokma, and Martin Hetherington, *Ecosystem Analysis,* Michigan State University Extension Bulletin (East Lansing, Mich.: 1973). Used by permission.

KEY:
I—EXCELLENT III—FAIR
II—GOOD U—UNSUITABLE

CONCLUSION

We are just beginning to appreciate the long-term and complex relationships between soil, water, vegetation, and cultural environments. Because of the severe soil-related problems that can result during and after construction on a site, many states and communities have set standards limiting and guiding new developments. In most cases the services of expert *hydrologists* (water specialist) and *pedologists* (soil scientists) are necessary for reliable interpretation and forecasting of long-term effects. Closely related to these specialized fields is *limnology*, the study of fresh-water ecology. The relationship of fresh water to planning will be covered in the next two chapters.

REFERENCES

Foth, Henry D. *A Study of Soil Science*. Chesterton, Md.: LaMotte Chemical Products, 1970.

George, Robert W., Mokma, Arnold, and Hetherington, Martin. *Ecosystem Analysis*. Michigan State University Extension Bulletin. East Lansing, Mich.: 1973.

Kohnke, Helmut. *Soil Science Simplified*. Lafayette, Ind.: Balt, 1966.

Lotkowski, Wladyslaw M. *The Soil*. Chicago: Educational Methods, 1966.

Mokma, A. A., Derch, E. and Schaner, D. J. *Land Judging in Michigan*. Michigan State University Cooperative Extension Bulletin, No. E-326. East Lansing, Mich.: 1974.

U.S. Department of Agriculture, Conservation Service. *Environmental Do's and Don't's on Construction Sites*. Misc. Publication No. 1291. Washington, D.C.: 1974.

U.S. Department of Agriculture, Soil Conservation Service. *Know the Soil You Build On*, by A. Klingebiel. Bulletin No. 320. Washington, D.C.: 1967.

———. *Teaching Soil and Water Conservation—A Classroom and Field Guide*, by Albert B. Foster and Adrian C. Fox. PA-341. Washington, D.C.: 1957.

Whiteside, E. P., Schneider, I. F., and Cook, R. L. *Soils of Michigan*. Michigan State University Cooperative Extension Bulletin, No. E-360. East Lansing, Mich.: 1968.

9 WATER AS AN ECOSYSTEM

It should be emphasized that a knowledge of fresh-water ecology is very important to planners, for as water moves across the surface of landscapes, it adds another dimension to our study. We must look at water as a dynamic, ever-moving force respecting no boundaries and sustaining life on every inch of ground it covers. In a sense, it is the lifeblood of all communities, the element that weaves together the fabric of the total ecological system.

If you are taking or have taken science courses in physics, chemistry, and/or biology, much of the following will be familiar to you, but it is basic to understanding social problems of land-use planning. If the natural sciences are not in your course of study, now is your chance to learn a little more about your physical, chemical, and biological world.

Some major problems that result from poor land development are directly concerned with water. The analysis of these problems can be a very difficult task. Water can cause flooding problems that are aggravated by accelerated runoff from urban areas. Land-use changes can cause rooted plants and algae to grow at a faster than normal pace, causing lakes to become weed-choked. Construction can disrupt the normal flow of runoff, resulting in a variety of undesirable conditions ranging from flooding and sedimentation to the destruction of fragile habitats. Water problems are truly massive. Many of these problems are caused by our lack of understanding of the physical characteristics of water and how these physical characteristics relate to plants, animals, soils, and engineering practices.

WATER AS A UNIVERSAL SOLVENT

Water has several unique properties that must be understood in order to predict the effects a development will have on a body of water. One of these properties is its ability to dissolve other substances. It has often been called a *universal solvent*. It dissolves and disperses through its own substance the molecules of many mineral nutrients and gases, thereby making these substances available to the variety of life in the water. It might seem that aquatic plant growth would be unlimited in such an environment, but the relative concentrations of miner-

FIGURE 9.1. What once was the edge of a farm pond is now a flow of mud caused by erosion.

als and gases and the rate at which these substances disperse through water can limit the growth of aquatic organisms. Substances disperse at different rates in water than they do in air. Water, for example, contains thirty times less oxygen for a given volume than does air. Nitrogen occurs in very small quantities in water; carbon dioxide in water and air is very similar in concentration.

The rate of dispersion of oxygen into water can be changed by varying the temperature. In turn, a slight variation in the concentration of oxygen can have a radical effect on organisms in the water. For example, man's activities can cause the water to warm. Warm water holds much less oxygen than cold water. Animals may then die for lack of it.

Or a stream may become muddied from runoff from a farmer's field. Plants are unable to photosynthesize in dirty water; their absence decreases the oxygen and increases the carbon dioxide. Animals can also limit the concentration of materials in water. Protozoa, worms, insect larvae, bacteria, and other small organisms act as decomposers; they feed on organic substances in the water and use up oxygen in respiration, thereby reducing the amount of available oxygen for other wildlife forms.

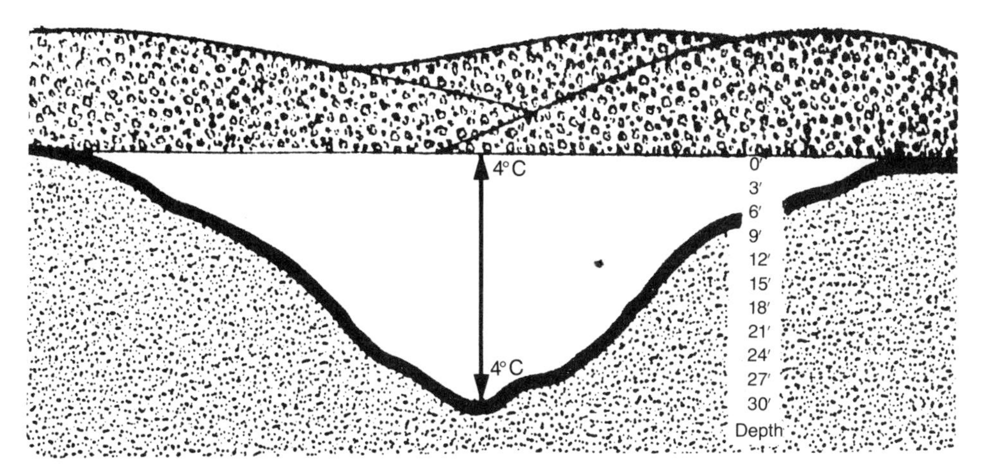

FIGURE 9.2. In the spring and fall, the temperature is the same throughout the lake, and nutrients and gases are mixed by the wind.

TEMPERATURE-DENSITY RELATIONSHIPS

Another unique characteristic of water is its temperature-density relationship. Water can store a large amount of heat. It is also slow to give up heat. A small pond warms or cools very slowly compared with the land. Water is also most dense at 4° centigrade and is less dense when a gas or a solid. These factors can combine to create a variety of living conditions in deeper (twenty feet or more) lakes. The data collected by a researcher may be dramatically different at varying depths in a lake and at different times of the year.

In winter, the cold water (0° centigrade) is found at the surface. The warmer but heavier water (between 1° and 4° centigrade) sinks to the bottom. As spring comes and the sun warms the surface, the temperature of the surface rises to 4° centigrade like the rest of the lake. At this point the temperature is the same throughout the lake and the density of the water is the same. Now the wind can mix the water freely. Nutrients and gases can be distributed equally throughout the volume of water by the currents. This period is often referred to as the *spring overturn*. (See Figure 9.2.)

During the summer, the surface of the lake heats more rapidly than the water below, so a wide variation in temperature results. Resistance (*viscosity*) is created between the two layers, preventing their mixing. As a result, the surface waters remain livable because of their exposure to the oxygen in the air and the mixing action of the wind. This surface zone is called the *epilimnion*. Below the epilimnion is a relatively thin layer called the *thermocline*, where the water is dense and the temperature drops rapidly. Below the thermocline is the *hypolimnion*. Water remains "trapped" in this region. It may warm up a little, but it becomes depleted of oxygen because of the decomposition by bacteria of organic material in the bottom muck. This phenomenon is known as *lake layering* or *thermal stratification*. (See Figure 9.3.)

In the fall, the lake surface cools with the falling air temperature until it reaches a maximum density at 4° centigrade at the surface. The heavy water

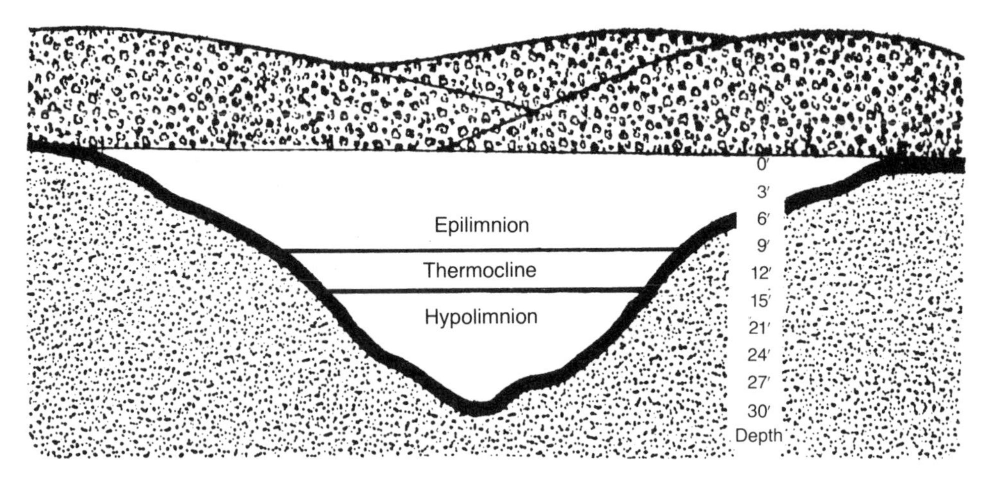

Epilimnion = high in temperature and oxygen
Thermocline = rapid drop in temperature and oxygen
Hypolimnion = low temperature and oxygen

FIGURE 9.3. In the summer, temperature variations cause the lake to stratify, resulting in poor circulation of nutrients and gases.

sinks, assisted by fall winds, and the lake once again has equal amounts of nutrients and gases throughout, in the *fall overturn*.

ECOLOGICAL CHARACTERISTICS OF WATER

In a sense, water functions within the natural system much as does blood in our bodies. The human body is a unit made up of millions of cells working together as tissues, organs, and systems. Using this analogy, we can define an *ecosystem* as a unit made up of millions of organisms all working together to form a single unit-system, such as a field or a pond. Within this unit, water functions as man's circulatory system does, bringing gases and nutrients to the many organisms and also removing their waste products. This is especially true in an aquatic system, where raw materials are cycled and recycled from plants to animals.

Photosynthesis and Respiration

The processes of photosynthesis and respiration can serve to illustrate the role water plays in cycling nutrients and gases. (See Figure 9.4.) Just as food gives us energy to carry out our daily tasks, the sun supplies the energy for work in the aquatic system. Microscopic plants (*phytoplankton*) and macroscopic rooted and floating plants are able to lock up the sun's energy by changing carbon dioxide and water into sugar molecules in photosynthesis. Green plants are called *producers* because they are the only organisms capable of making food by this process.

Once the food is made by the plants, it is sent to all parts of its body, where the food molecules are broken down and the energy released to reproduce and build tissues. This process requires additional material, such as proteins and minerals, which the plant acquires through other cycles. The process of extracting the energy from the sugar molecule is called *respiration*. During this process, oxygen is used, and the carbon dioxide and water return to their original form. Almost all organisms, plant and animal, use oxygen during respiration, whether they live on land or in water.

Food Relationships

Animals do not have the ability to capture the sun's energy, so they must rely on plants for their energy source. In ecological terms they are known as *consumers*.

FIGURE 9.4. Diagram of the energy cycle.

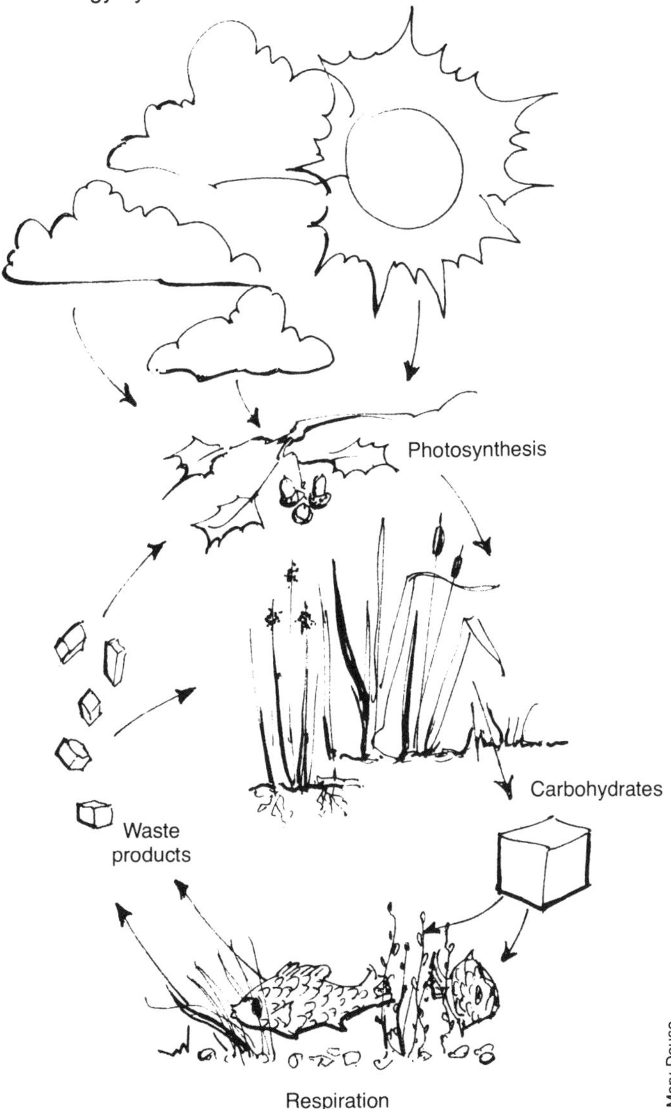

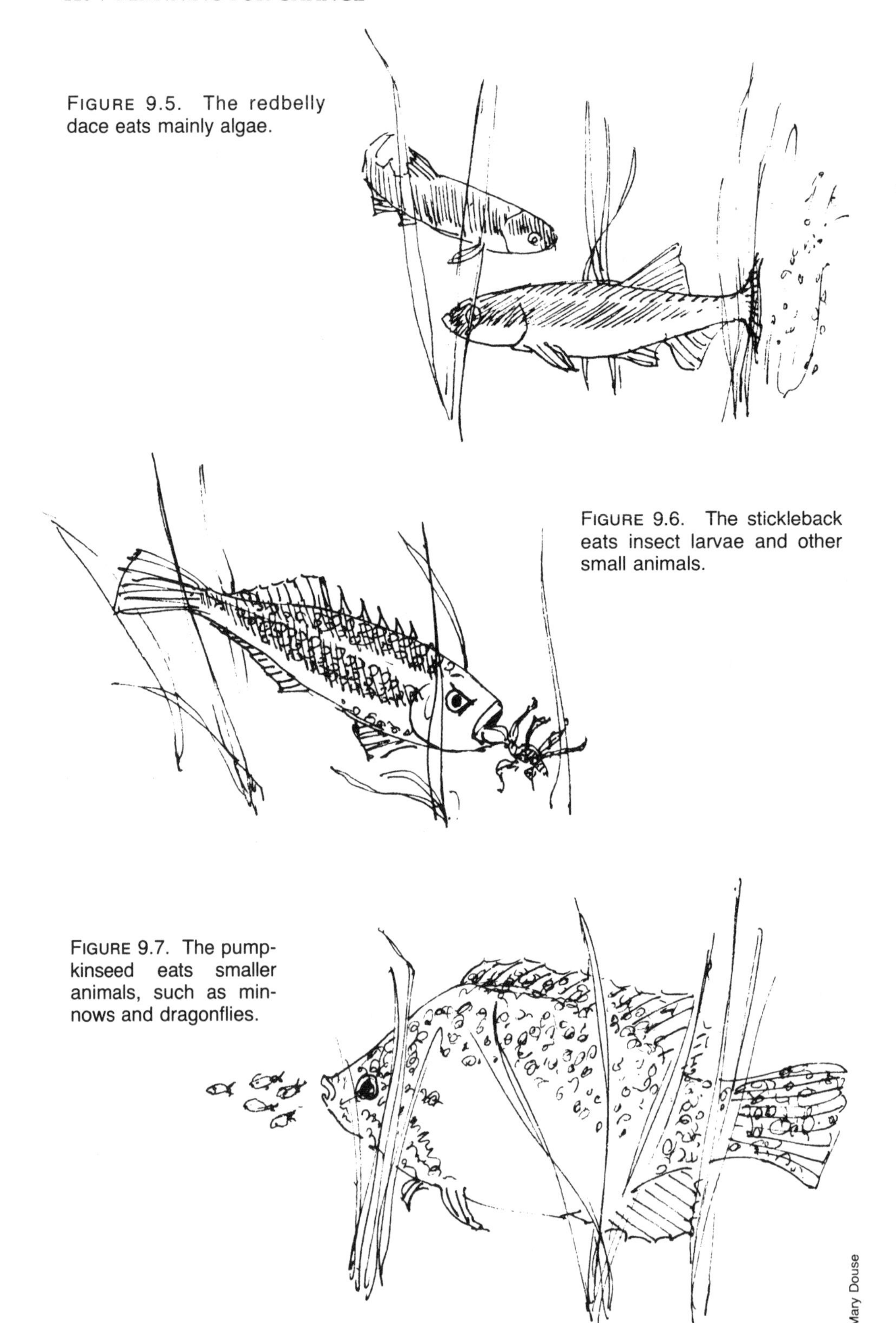

FIGURE 9.5. The redbelly dace eats mainly algae.

FIGURE 9.6. The stickleback eats insect larvae and other small animals.

FIGURE 9.7. The pumpkinseed eats smaller animals, such as minnows and dragonflies.

Some animals feed primarily on plant life and are known as *primary consumers*. On land many insects and birds are primary consumers, as are such mammals as mice, rabbits, and deer. In the aquatic environment, microscopic animals known as *zooplankton*, insects, and some larger animals serve in this role.

These small animals living in the water may be eaten by insect larvae or small fish. The small fish and insect larvae may, in turn, be eaten by larger fish or insects. The animals that feed on the primary consumers are called *secondary consumers*. In the end, the body materials of all organisms are broken down by *decomposers* such as bacteria and *scavengers* (crayfish, worms, clams, and so forth) into simple elements. These elements are then used by plants to start the cycle over again. This process is called the *food web*.

In the food web, energy, gases, and nutrients flow from plants to animals. Under normal conditions, a *dynamic balance* is established. Any fluctuation in the normal source of energy, gases, and nutrients required for life can have an effect upon individual species within the community.

On grassy lawns and on farms, we encourage the increased productivity of one species of plant over another. We tip the balance toward favored plants, partly by adding fertilizers (additional nutrients) to the soil. When extra nutrients are added to a lake, the balance between producers and consumers may be upset. For example, the amount of phosphates and nitrates needed for plant growth is extremely small; under some circumstances, a small amount of phosphates (from fertilizers, detergents, a feedlot, or a sewage-treatment plant) can supply the needs of thousands of pounds of algae, thereby resulting in an "algal bloom" or an "explosion" of algae in a lake.

Lake Aging

Every body of water is subject to a natural aging process. If left undisturbed, a body of water undergoes changes. This is true of small ponds and large lakes. Generally, three types of natural changes can occur: (1) the lake is gradually filled in with eroded soils and organic substances from the surrounding drainage basin; (2) dissolved and suspended materials carried into the lake encourage increased production of plants and animals, which in turn die and add their remains to the bottom; (3) water-loving plants at the lake's edge create sedimentation and other conditions, which move the shore's edge farther out into the lake. These forces cause the lake to become smaller and shallower. Eventually, the lake fills in completely.

As the natural aging continues, the types of plants and animals change. Young (*oligotrophic*) lakes are deep and cold, low in nutrients and high in dissolved oxygen. Desirable game fish such as trout, bass, pike, and pan fish may be plentiful. An older (*eutrophic*) lake has more nutrients, is shallower, heats more rapidly, and has less oxygen. Rooted plants occupy much of the lake's surface. Game fish that are found in cooler lakes are replaced by such so-called rough species as carp, suckers, and bullheads. (See Figures 9.8 and 9.9.)

Succession in a Lake

The rate at which a lake becomes eutrophic depends on such factors as its temperature, size, and depth. As an oligotrophic lake becomes eutrophic, cer-

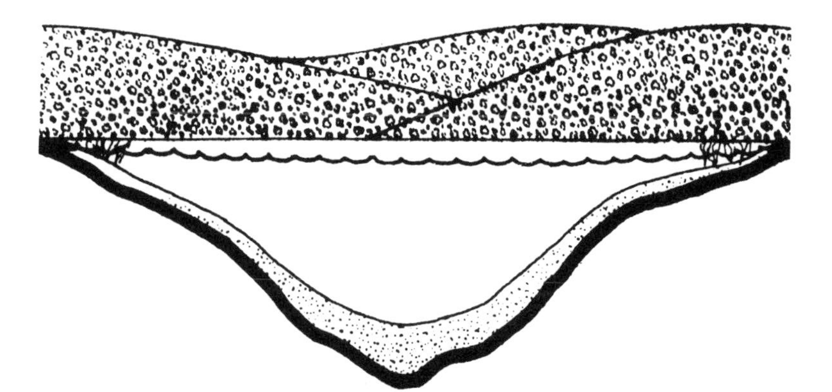

FIGURE 9.8. Oligotrophic lake: cold water, high in oxygen, low in nutrients.

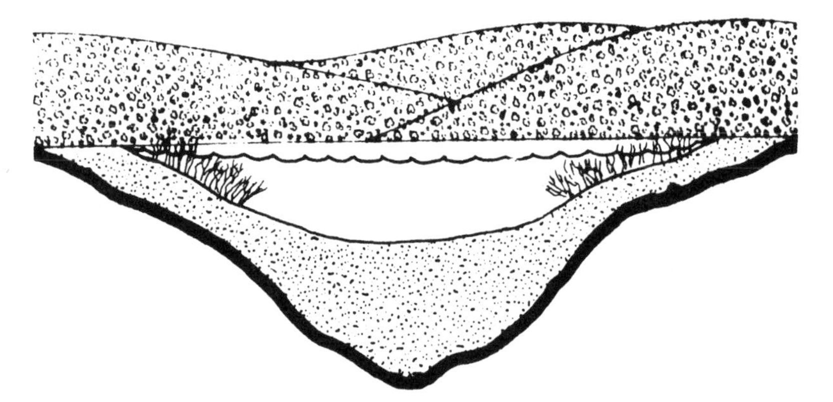

FIGURE 9.9. Eutrophic lake: higher temperature, lower in oxygen, more nutrients, shallower.

tain predictable growth patterns occur. Zones along the edge of the lake are established one after another, each with its representative group of plants and animals, in a manner similar to the succession by different plant-animal communities on land. Figure 9.10 shows the types of zones that may dominate a lake in north central and northeastern states.

Like the vegetative communities found on land, the aquatic plant zones are mostly labeled by their visually dominant plants. The *lake's edge* (shore) is the zone familiar to most people. Here the sun- and water-loving trees grow, along with a variety of animals that use the area for nesting, cover, and food.

In the shallow water near the shoreline, such *emergent plants (aquatics)* as cattails, reeds, rushes, and sedges dominate. Microscopic organisms, crustaceans, and insects are dominant forms of animal life. Because of the great variety of food and protection offered by the plants, many large predators use the area as a nursery for their offspring.

In somewhat deeper water, scattered mats of *floating plants (aquatics)* such as pond lilies and watershield, can often be found. These plants are anchored to the bottom by long, slender stems. The larger specimens of fish can often be found among these stems.

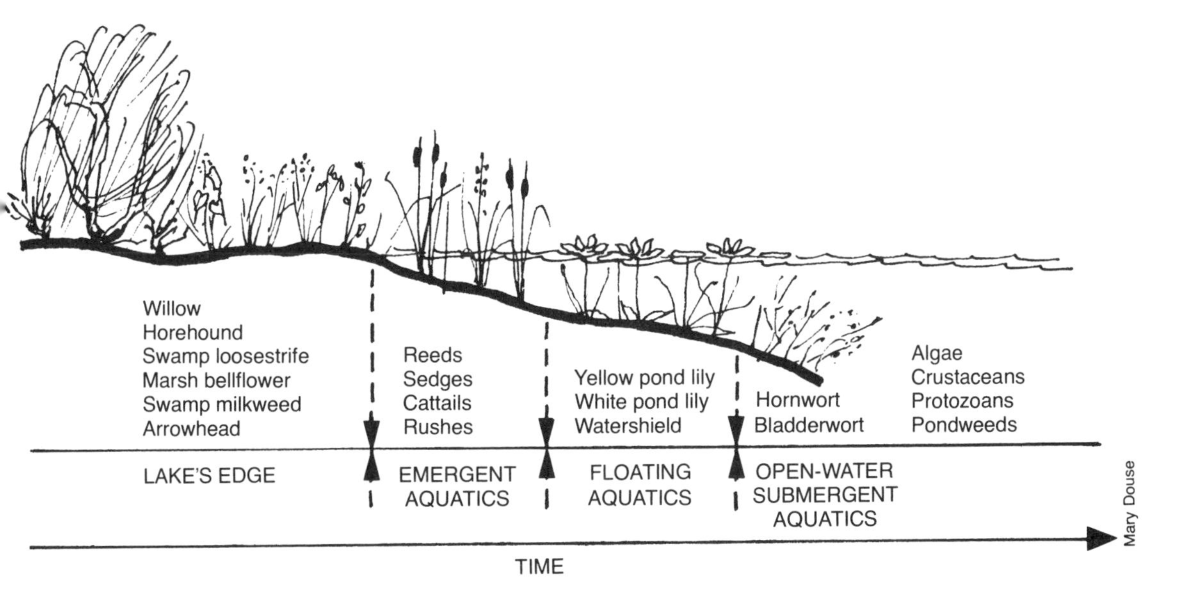

FIGURE 9.10. Profile of major zones of plant growth and their dominant plants.

Beyond the water lilies lies open water, where free-floating, microscopic organisms abound. Algae, protozoans, small crustaceans, and rotifers are common. Below the surface, rooted plants capable of life underwater may exist where the water is not too deep. These *open water submergent plants (aquatics)* can be very common in the open water, as well as in the emergent and floating zones. Many of the submergent plants are important sources of food for

FIGURE 9.11. Water lily.

wildlife. They also can be a major cause of water pollution, causing the lake to become choked with vegetation and thus preventing boating and swimming.

As a lake ages, the zones move toward the center or the deepest portion of the lake until the lake turns into a marsh, bog, or swamp. This is a normal process but people unfortunately can speed up the eutrophication by increasing the amount of suspended and dissolved materials that enter the water, for instance by sedimentation and fertilization.

Aquatic Organisms

Throughout the vegetative zones, representatives of almost all forms of life can be found. These organisms range from the microscopic bacteria, protozoans, and algae to fish, turtles, snakes, and even large trees.

Except for fishermen and naturalists, most people in the past had little knowledge of the animals and plants that live in water. With the influx of people to the urban fringes, lake living is now common. This has led to problems, the most severe being the competition for water space between aquatic plants and people and the pollution of what used to be potable water.

As a result, the public is becoming more aware of a greater variety of aquatic plants and animals and the ecological principles that govern their life styles. The many controversies surrounding lake living have focused attention on such organisms and animals as algae, water weeds, and carp. Attempts to eradicate troublesome organisms are varied, often fruitless and costly.

The public is also slowly becoming aware that troublesome organisms are only symptoms of the larger problem of eutrophication, and that the problem animals and plants are only indicators of certain physical and chemical conditions that exist in the lake. Limnologists have used plants and animals as indicators of aquatic environments for a long time; now a study of these organisms can be helpful to planners also.

In the last section of the Activities Manual that accompanies this text, you will find a series of illustrations that show some common animals and plants. Perhaps some of these occur at a site where you are planning for development. If so, the illustrations may be helpful in identification. A picture of each organism serves as a key to identification. Accompanying the picture is a simple description of the scientific family or other groups (taxa) to which the organism belongs. Some of the organisms are indictors of water quality or water type. The list is only a simple guide. More detailed classification and descriptions should be obtained from books with biological keys.

CONCLUSION

The physical characteristics of an aquatic environment are modified over time by changes in the surrounding landscape. Some of these changes are seasonal; others occur as a natural process of lake aging. A knowledge of these natural occurrences can help you determine the conditions that exist in a given body of water prior to any proposed changes to the surrounding landscape. These data can then be compared to data obtained from studies carried out after development, which can serve as evidence of accelerated aging or pollution. Chapter 10 discusses some of the techniques that can be used to study, record, and illustrate the physical, chemical, and biological "health" of a body of water.

REFERENCES

Andrews, William A., ed. *A Guide to Freshwater Ecology*. Englewood Cliffs, N.J.: Prentice-Hall, 1972.

Buller, David. *Pond Guide*. Berkeley, Calif.: University of California, Outdoor Biology Instructional Strategies, 1975.

Fulton, Jerome K., Say, Wayne E., Miller, Warren P., Bletcher, Thomas E., and Koch, Henry C. *Inland Lakes—Analysis and Action*. Michigan State University Cooperative Extension Bulletin, No. E-718. East Lansing, Mich.: 1971.

League of Women Voters Education Fund. *The Big Water Fight*. Brattleboro, Vt.: Stephen Greene Press, 1966.

Michigan Department of Conservation. *Aquatic Weeds and Their Control in Michigan*. East Lansing, Mich.: 1964.

Renn, Charles E. *Our Environment Battles Water Pollution*. Chestertown, Md.: LaMotte Chemical Products, 1969.

U.S. Department of Health, Education, and Welfare. *Limnological Aspects of Recreational Lakes*, by Denneth M. Mackenthun, William M. Ingrams, and Ralph Porges. Washington, D.C.: 1964.

Oakland County Planning Commission

10

WATER: INVENTORY AND ANALYSIS

Virtually all water-connected urbanization problems discussed in this text can be grouped into four categories: (1) increased volume of runoff water; (2) increased speed of runoff water, (3) increased amounts of dissolved and suspended materials in a body of water, and (4) sedimentation where sediment-laden water slows down. This means that the planner or site investigator must collect data indicating the existing condition of any bodies of water, determine the potential for runoff problems that may result from a development, and try to predict what the future will bring to the water system.

Some of the data collected about a water body can serve as *base-line data*. Base-line data describe an existing situation and serve as a base for comparison with later data. Later comparison may suggest that water quality has deteriorated or improved, or that the volume and rate of runoff from surrounding areas has increased or decreased.

With water data, recommendations can be made governing construction practices that may help alleviate future water problems and make the wisest possible use of water bodies and surface runoff. Many states have regulations controlling the permissible degree of erosion and runoff, but too little has been done to regulate water quality during land-use changes. This chapter will discuss how to analyze surface water and how to illustrate pertinent data.

RUNOFF OF SURFACE WATER

There are three basic types of water that a planner should be familiar with when studying a site: (1) *runoff*, which is drainage of surface water across the landscape; (2) *streams*; and (3) *seas*, *lakes*, or *ponds*. Marshes, bogs, and swamps (which were covered in chapter 7) should also be considered part of a water study. Groundwater (dealt with briefly in chapter 7) is of course importantly related to the other types.

Drainage Basins

Water is subject to the force of gravity. It always moves downward across the landscape, leaving behind distinct drainage patterns called *drainage basins*, or

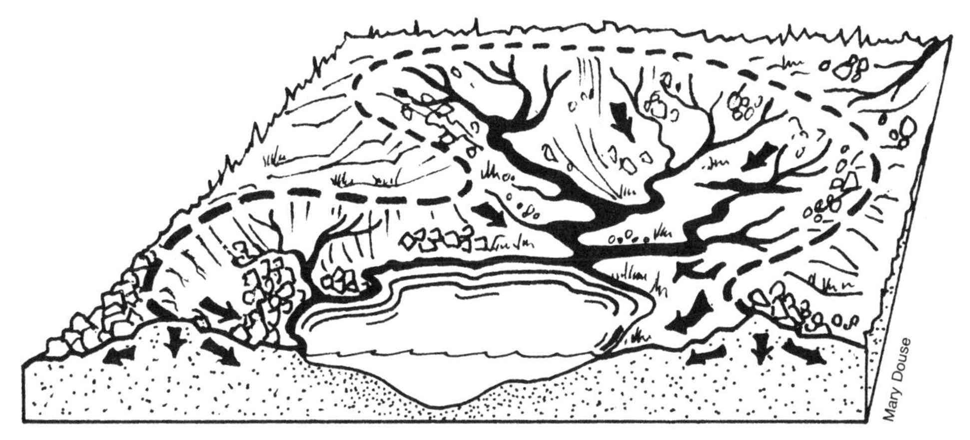

FIGURE 10.1. Drainage basin (watershed) showing divide and direction of water flow (note arrows).

watersheds. The term implies a valley, the upper edge of which is called a *divide*. (See Figure 10.1.)

Groundwater

The precipitation not utilized by plants becomes either surface water or groundwater. If the soil permits, some of the water percolates down through the soil until it reaches the water table. Gravity continues to influence the groundwater and to move it through the soil, usually in the direction of the slope. If the land slopes, a certain percentage of water runs downhill off the surface of the ground and settles in low spots, forming puddles, ponds, lakes, or oceans (except in arid regions where the water may evaporate before ponding). (See Figure 10.2.)

Any system of runoff can be disrupted by development. Even a nature trail built along a causeway through a swamp can upset the natural seepage of water

FIGURE 10.2. Movement of water in the environment.

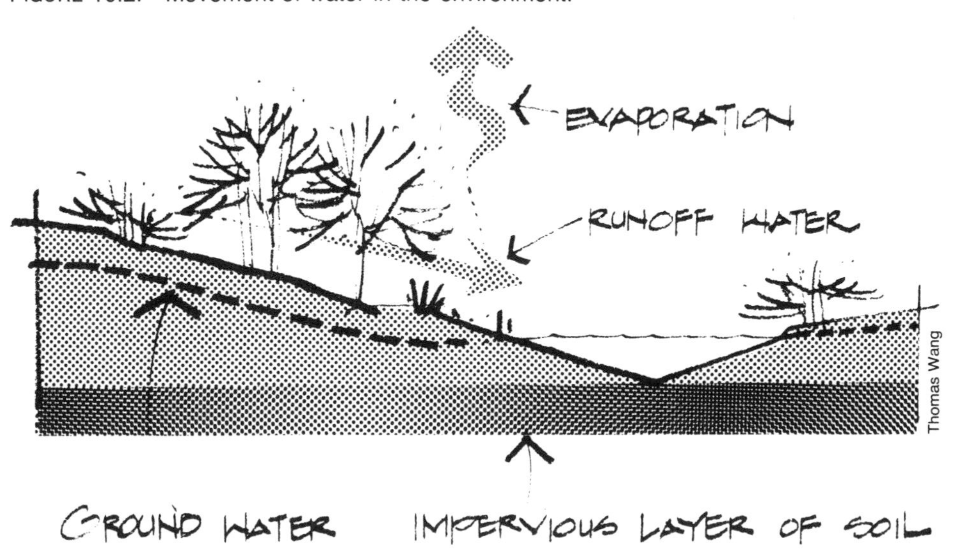

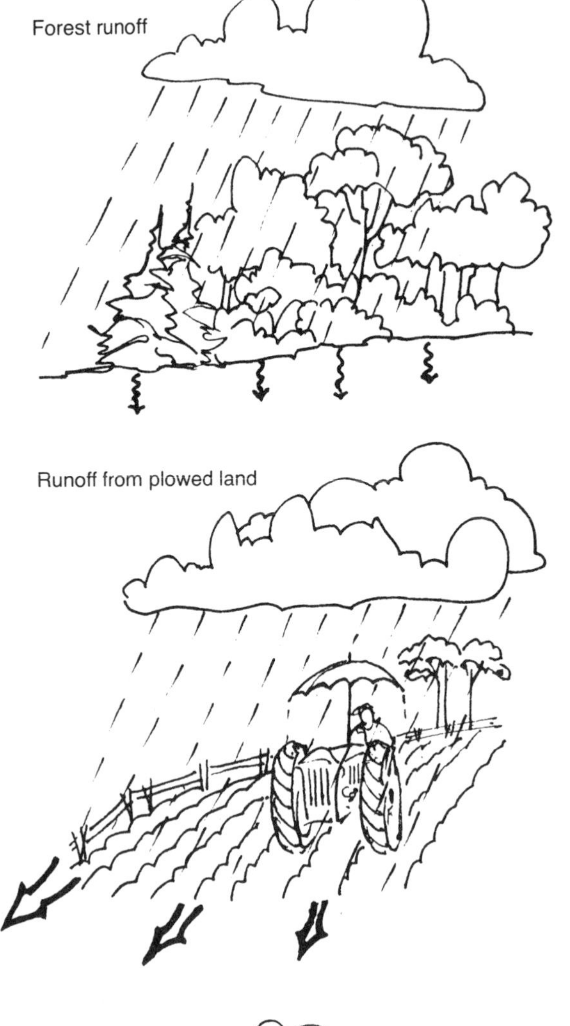

Mary Douse

FIGURE 10.3. Forested land protects the soil environment from surface water runoff and maintains adequate water flow into streams and groundwater. Removal of the vegetation during agricultural and urbanization practices causes the water to run off the land at a more rapid rate and in greater volume.

by acting as a dam. The water may rise on one side of the trail, while on the other side it remains lower. This modification can affect the plant communities drastically unless artificial drains are planned to alleviate the condition.

By carefully analyzing and describing a drainage pattern, an engineer can plan construction so that a minimum of the natural water flow will be disturbed. The rate at which water runs off the land is proportional to the slope of the land, the amount of vegetative cover present, and the structure of the soil. Given the same slope, precipitation will run off urban areas and cultivated areas much more rapidly than off a forested area. As a rule, forest land will absorb 80 percent to 100 percent of the water that reaches the forest floor. In the soil, the water will remain cool; gravity will move the water slowly into a nearby stream, thereby maintaining a relatively constant stream-flow throughout the year. Removing trees and other ground cover causes more rapid runoff, and substituting asphalt and cement for porous soil and vegetation results in little or no penetration of water into the soil, with consequent very rapid runoff. (See Figure 10.3.)

ANALYZING AND ILLUSTRATING RUNOFF

To understand the drainage patterns that exist on a site, you as a planner must sometimes isolate the various drainage basins. Contour maps can supply most of the information needed for determining the major drainage patterns.

Figure 10.4 illustrates a number of major drainage patterns identified on a given site. If you examine the drawing carefully, you will notice that the divides that separate the basins have one common characteristic: they run along the tops of a number of hills and major ridges on the site.

From such a map, engineers can determine the most feasible locations for roads and buildings. Figure 10.5 shows the areas that must be cut and filled in order to build the roads in the development. Such a map utilizes the drainage patterns as much as possible in order to minimize the effect that such disturbances will have on the environment.

STREAMS

Eventually, much of the water that runs off the surface of the land and through the soil will enter a stream. Whether a moving body of water is labeled a stream, brook, creek, drain, or river often depends on the history of the body of water and the area of the country in which it is found. (The term *stream* is a general one used by hydrologists to describe any water flowing through any channel, even a pipe or garden hose.

Intermittent Streams

Every large river has countless miles of small contributary streams that flow only when the volume of runoff water is high. These are known as *intermittent streams*. Unfortunately, because they lack water during drier portions of the year, they are often considered a nuisance and filled in with soil. Few laws cover the important role that intermittent streams play in moving water into a river system.

(continued on page 123)

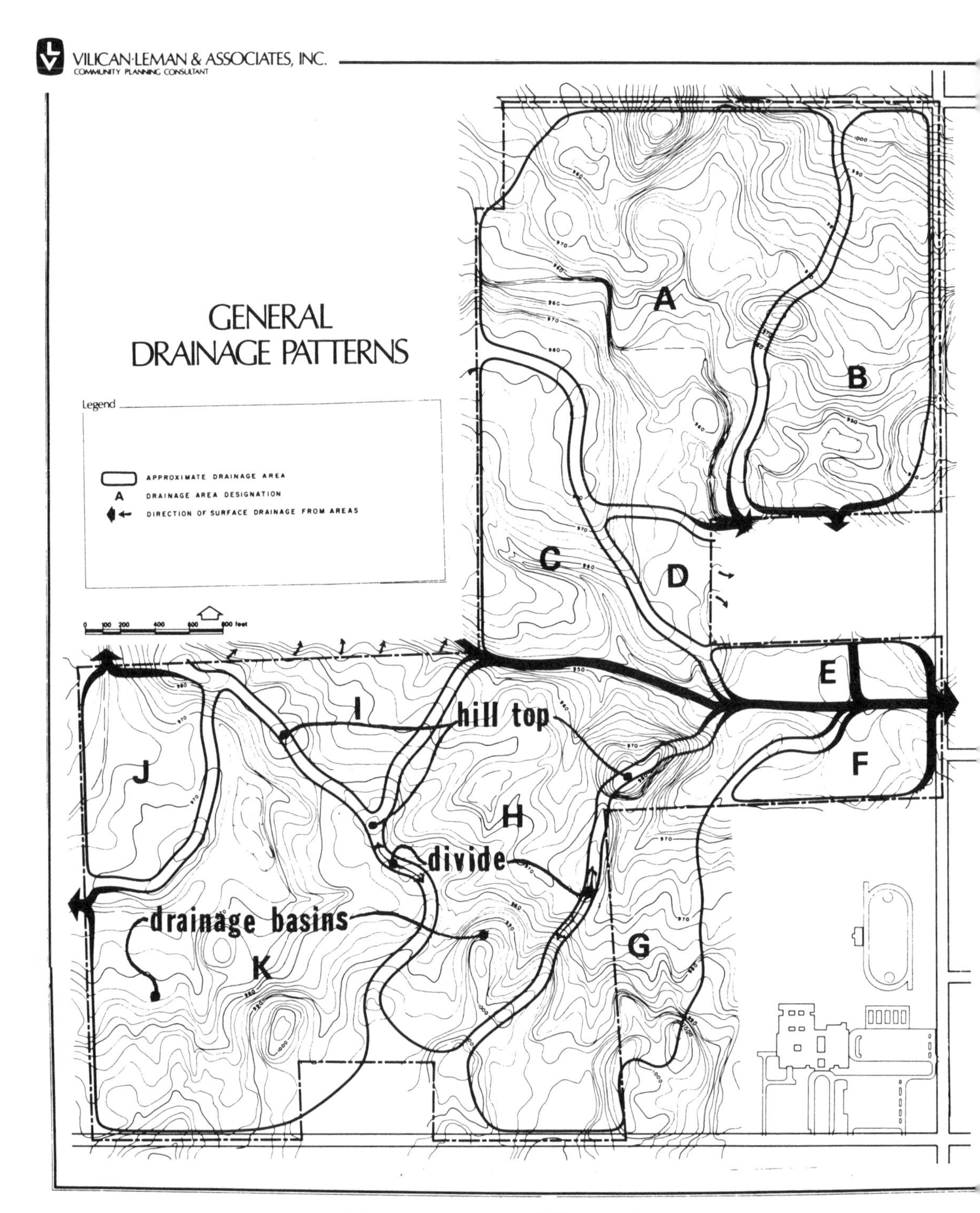

FIGURE 10.4. Major drainage patterns on a 377-acre site.

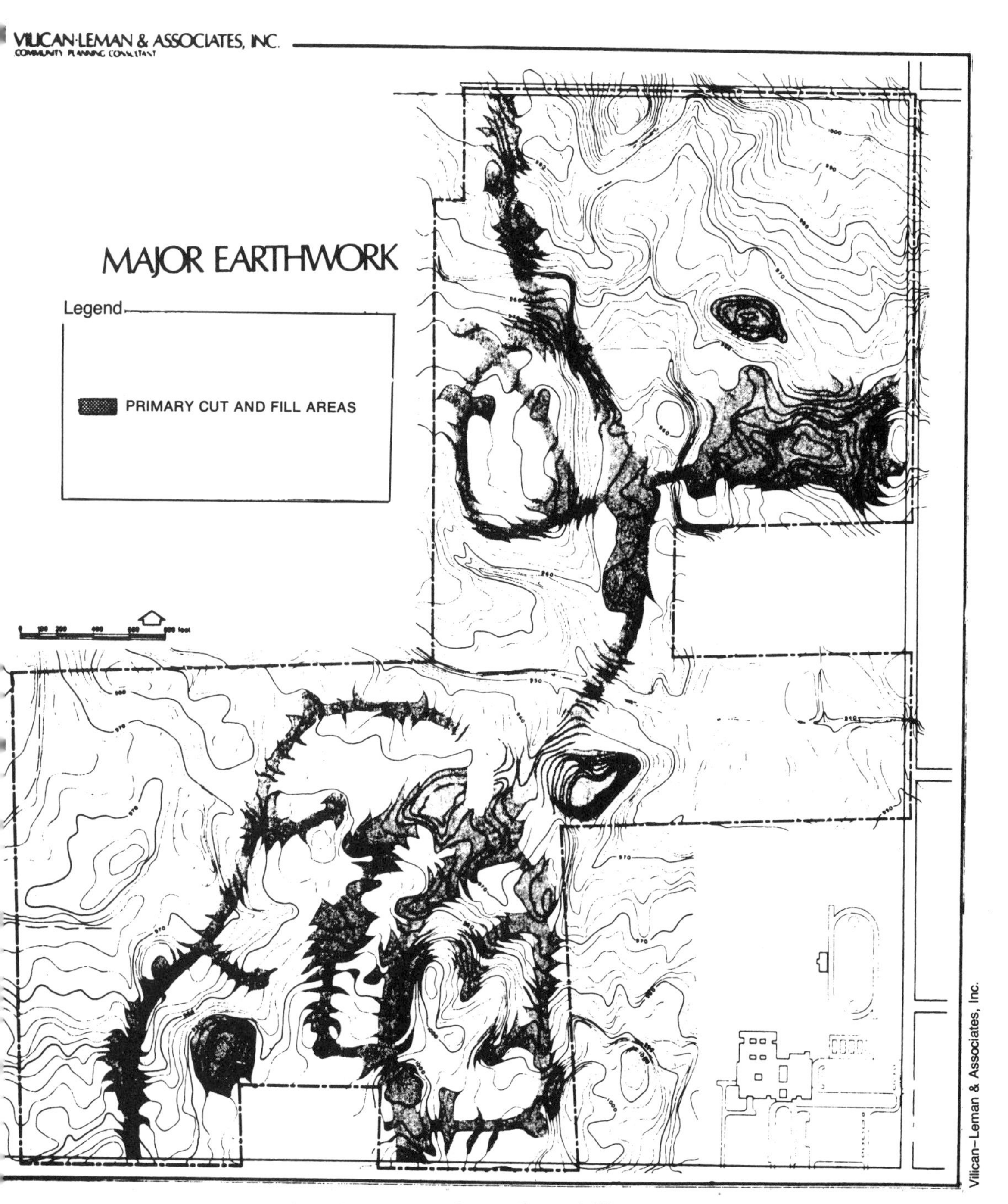

FIGURE 10.5. Map showing primary areas for cutting and filling.

FIGURE 10.6. Intermittent stream.

FIGURE 10.7. Relatively natural drain.

Dave Larwa

FIGURE 10.8. Concrete drain.

Dave Larwa

Drains

The term *drains* pertains to man-made stream systems. The term is sometimes confusing, because natural streams are often modified to some degree and are then classified as drains. In some states, a stream is classified as a drain when a county drain commission or city engineering department is directed by law to control the volume of flow or maintain other features that were built to facilitate stream flow.

Traditionally, drains have been constructed to remove, by artificial means, excess water from an area in order to improve its use for agriculture, building, or mosquito control. Today over a million acres of land have been drained for these purposes. Approximately 30 percent of this acreage is in Indiana, Michigan, Minnesota, Ohio, Illinois, and Iowa, with Florida, Louisiana, Texas, and Arkansas each accounting for approximately 5 percent of the total. California leads all the western states in acres of both drainage and irrigated land.

The problems that can result from draining wetlands can be serious. The most common of these are flooding, fire, wind erosion, loss of wildlife habitat, rapid decomposition of organic matter in the soil, and the lowering of the water table.

Streams and Flood Plains

Small streams and drains eventually collect into larger and larger bodies of moving water (except in arid regions). These bodies of flowing water and their

FIGURE 10.9. Stream.

adjacent land areas are important natural features in any community. A stream's flood plain is that area of land abutting the stream that is subject to flooding during heavy rains and snowmelt, with runoff greater than its channel can carry. Flood plains are often the subject of controversy because of problems that result when the plains are built on and then flooded.

ANALYZING AND ILLUSTRATING DATA ABOUT STREAMS AND FLOOD PLAINS

The cross-sectional profile of a flood plain varies with the size of the stream and the nature of the area through which it moves. Most streams in urban areas have a profile showing high ground, a more or less terraced slope leading down to a flat river-bottomland (flood plain), and a stream channel. (See Figure 10.10.) In the flood plain, normal or exceptional floodwaters move and deposit layers of sediment. This occasional deposition maintains the flat character of the flood plain, with its fertile silts well suited for agriculture, forestry, and wildlife. Flood plains also serve admirably for kinds of recreation that do not require structures.

Because of flooding problems associated with streams, you as a planner should identify any flood-prone areas on your proposed development site. Most

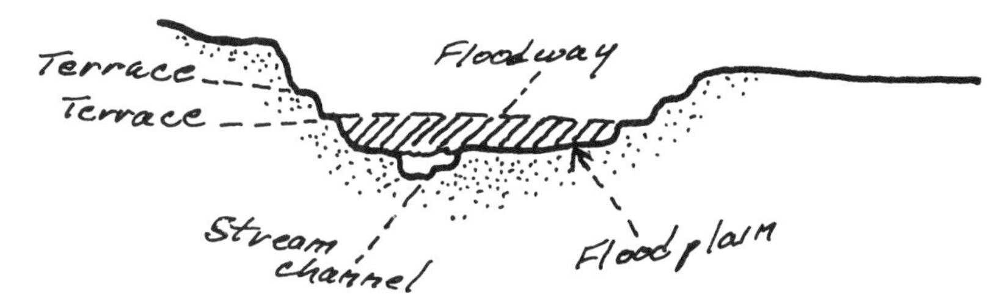

FIGURE 10.10. Cross-sectional profile of a mature stream.

city ordinances restrict development below the 2 percent flood level. This level, formerly and misleadingly referred to as the fifty-year flood level, is the height to which a river may rise twice in a hundred years (possibly two years in a row).

The determination of the 2 percent flood level requires engineering studies that can be very costly. Restrictions in most areas are therefore based on an arbitrary distance, such as one hundred to five hundred vertical feet above the stream bed. Such a distance is convenient for writing broad legislation to cover large municipal boundaries but is too vague to be accurate for small developments. As a result, stream floodways should be determined separately for each project.

What can you do to identify the width of a flood plain and estimate extreme high-water levels? A topographic map will help you determine the width of a flood plain. The shape of the contour lines that follow and then cross a stream will indicate the width of its flood plain. Study the contour lines along the edge of the stream. The bed of a stream drops from its headwaters toward its mouth and is always lower than the adjacent land. Contour lines represent this relationship. On both sides of a stream, a contour line projects back on itself upstream, crosses the stream, and turns downstream again. (See Figure 10.11.) This downstream bulge of the contour line as it crosses the stream is a fairly accurate measurement of the width of the stream. It will not show on all topographic maps, however, because the contour interval may be too large.

Stream Profile

One or more cross-sectional profiles of the stream itself should be included with a description of the flood plain. A profile can help to show the present condition of the stream bed and the composition of the stream bottom.

Rapid runoff of large volumes of water may destroy the character of a stream bed. The banks can be undercut, changing the course of the river and toppling

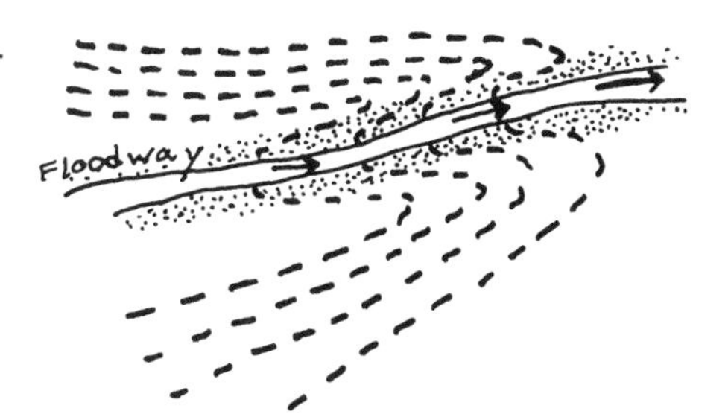

FIGURE 10.11. Width of a floodway.

streambank vegetation. Log jams from felled trees may result and act as dams restricting water flow. Also, sediments deposited on the stream bed from erosion upstream can change the velocity of the stream, affect the temperature of the water, and change the kinds of plants and animals that live in the stream.

A good picture of a stream can be obtained by making studies of it at random or at carefully selected points along its course in your site. Measurements of water depth can be taken with a metered stick or a weighted, metered string. The metered string can be suspended from a cord stretched across the stream between two stakes. The complexion of the bottom can be established at the same time by taking samples from the middle of the stream and from near both edges. (See Figure 10.12.)

Stream Volume

Knowing the volume of water a stream carries can help you determine how rapidly the water runs off the land upstream from your site. You can obtain the volume by determining the average size of the floodway. This is done by measuring the width and depth of the water in the channel at several random points along the stream. An average cross-sectional area is calculated from the width and depth measurements. The velocity of water is then obtained by recording the amount of time it takes for an object to travel one hundred meters (or feet) in the same area of the stream. Multiplication will then show the number of cubic units of water that flow past a point in a given period of time, as cubic feet per second (cfs) or cubic meters per second (cms).

Sedimentation Tests

Because runoff is always more rapid from cities than from well-vegetated rural areas, urban streams may carry a large amount of sediment. Many of the problems related to sedimentation (temperature and oxygen fluctuation, destruction of wildlife habitats, and so forth) have already been noted. You can conduct tests easily and with a minimum of equipment to determine the amount of sediment carried by a stream. Collect water samples before, during, and after a rainfall. After they have settled for a time, the amount of sediments can be compared.

If appropriate equipment is available, you can also test the amount of dissolved materials in the water, such as pesticides, salts, and oxygen. A major

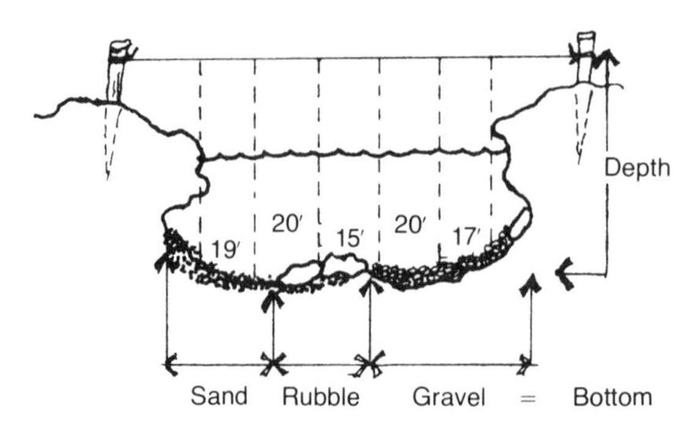

FIGURE 10.12. Sampling and measuring across a stream bed.

FIGURE 10.13. Diagram of sedimentation test.

concern of the land planner is sediments that come from stripped land during construction, so you should recommend in your proposal that test be done at that time. As in other studies, a number of tests should be made to obtain an average for the stream.

STANDING WATER

Standing water, as distinguished from streams, or flowing water, occurs in puddles, ponds, lakes and the like. Depending on the configuration of the land, runoff settles in low areas, where lakes, bogs, and marshes form. If the areas are low enough, groundwater may be exposed and may assist in the formation of the water body.

In urban areas, most lakes are small and are in some advanced state of succession. It is difficult to generalize about ponds and small lakes, for they all differ from one another, even though they may be close together. The nature of the drainage basin, amount of development, size, depth, and climate all influence the quality of the water.

Ponds

Intermittent ponds are small, shallow bodies of water caused by melting snow and/or seasonal rains. *Vernal pools* occur in spring, resulting from winter snows and/or rains; they dry up in summer. In semiarid and arid regions, shallow lakes called *playas* may form after the rare rainstorms. Any such standing water can be very useful for educational studies, recreation, and collecting basins and/or settling ponds for runoff water during and after construction.

Permanent ponds are smaller and shallower than lakes. Sometimes rooted, aquatic vegetation covers much or even all of the surface of such ponds. No temperature layering (thermal stratification) occurs except in deep ponds, for instance some in abandoned quarries.

FIGURE 10.14. Permanent pond.

John Victory

Dave Larwa

FIGURE 10.15. A lake—or a large pond.

Lakes

The terms *pond* and *lake* are used somewhat differently in different regions. What is a large pond to one person may be a small lake to another. Lakes are, by some people's definition, too deep for rooted aquatic plants to cover the entire surface. Because of their depth (often over twenty feet), temperature layering occurs in many lakes during the summer.

ANALYZING AND ILLUSTRATING STANDING WATER DATA

Analyzing a pond or lake can be very difficult. Collecting accurate water information often requires a variety of equipment, some of it expensive. Because of the complex relationships that exist in an aquatic environment, the science of analyzing and evaluating the data (limnology) is a highly specialized one. Whatever your resources, though, you should attempt to collect physical, chemical, and biotic data about any water-body at a site for which you are proposing changes.

A physical analysis of a water-body includes bottom and edge studies and measurements of temperature and oxygen profiles and degrees of turbidity. Volume and depth studies are important; to a large extent they control the concentration of nutrients and light available for plant growth.

Mapping the Edge of a Lake or Pond

Using aerial photographs is the easiest way to determine the shape of a lake or large pond. Comparing its overall shape over a period of years can be useful in judging the rate at which it is aging. If plant growth is rapid, the edge of the lake may "move" rapidly toward its center. For your project, you do not have time to watch a water-body for some years, but perhaps you can compare recent and older photographs.

If aerial photographs are not available, a lake's or pond's edge can be plotted with data collected at the site with a compass or transit to determine directions and measuring by pacing, measuring tape or stadia rod.

A base line should be established with a compass along one side of the lake. Then, at equal intervals along the base line, you or a helper should measure at right angles to the base line to the water's edge. The procedure can be continued around the lake. Once a base line is established, triangulation can also be used to locate reference points along the lake's edge. Back in the classroom, a map of the lake's edge can be accurately plotted, using the notes made at the site. (See Figure 10.16.)

Bottom Profiles and Plotting a Small Lake

Along with knowing what the shape of the lake's surface is like, it helps to know the shape of the bottom. Depth is a critical factor in controlling rooted, aquatic plants. Most aquatic weeds grow in less than fifteen feet of water. Once a shallow depth has developed in a lake, plant growth can be very rapid and clearing excess vegetation from the lake becomes much more costly. Getting lake profiles can be important for basic data. Later studies can then give an investigator a good idea of just how rapidly the lake is filling in. Bottom profiles also help determine the volume of a lake and assist in locating collection stations for other physical, chemical, and biotic studies.

In regions that are cold enough, collecting data for a profile of the lake bottom can best be carried out in winter when strong ice covers the lake. Hand- or gasoline-powered ice augers can be used to drill holes through the ice. These holes act as stations through which to drop a metered line to the bottom; the depths are then recorded on your site map. The stations should be at equal intervals along a line established with a compass. In areas where ice does not

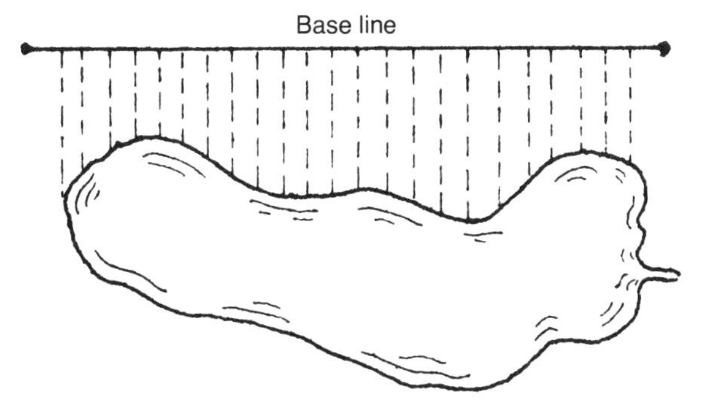

FIGURE 10.16. Mapping a pond or small lake.

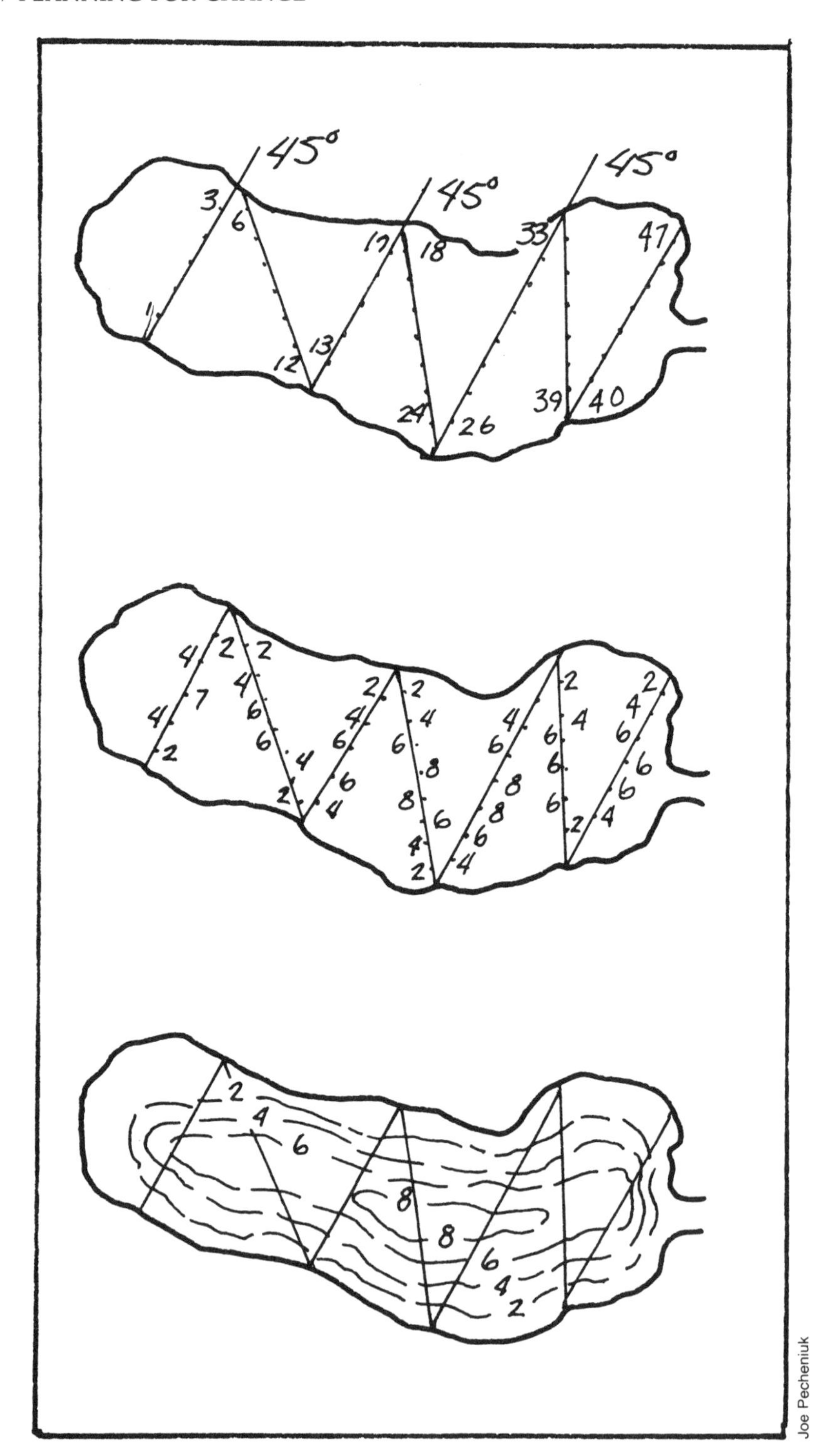

FIGURE 10.17. Mapping a lake bottom: stations where measurements are taken (top); measurements for each station (center); contour lines drawn from measurements (bottom).

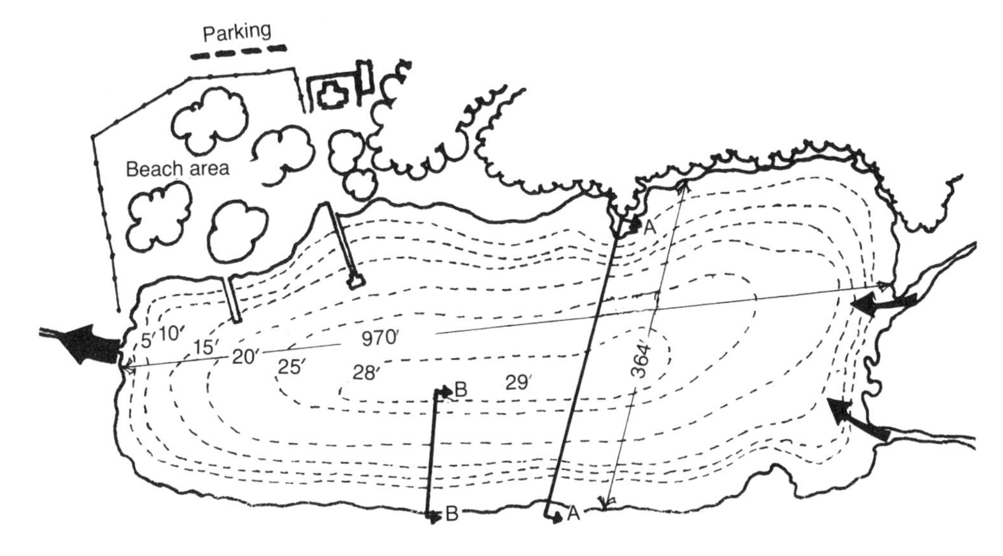

FIGURE 10.18. Map showing the shape and depth of a lake. Lines BB and AA refer to a cross-sectional or profile drawing of that portion of the lake. Figure 10.19 represents a profile view of section AA.

form, bottom profiles can be obtained with sonar devices used by fishermen or with a weighted, measured line that is lowered to the bottom from a boat crisscrossing the water systematically along compass bearings. (See Figure 10.17.)

Back in the classroom, contour lines are drawn to connect points of equal depth. A final map can then be drafted showing the physical attributes of the lake, with profiles to supplement it. (See Figures 10.18 and 10.19.)

Physical and Chemical Testing

Testing and analyzing water for its physical and chemical components can be both expensive and complex. For a land-use change, however, water testing can be less extensive. Recall that changes in land use usually result in problems related to accelerated runoff of increased volume and added dissolved and suspended materials. Impact on a water-body can be revealed by testing for several basic features, such as turbidity, major nutrients, temperature, and oxygen.

FIGURE 10.19. Profile of a lake bottom. (See Figure 10.18.)

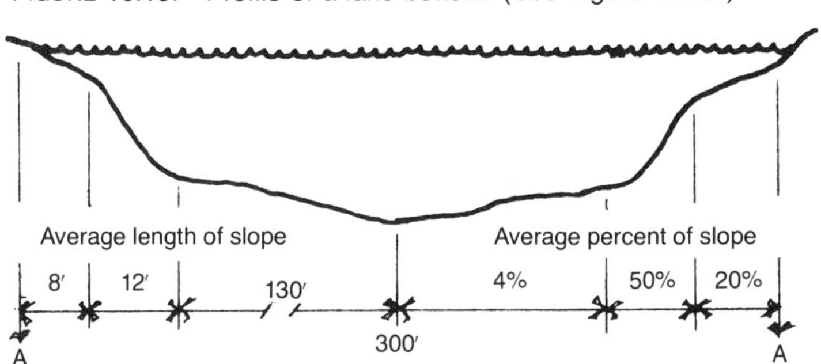

Dave Larwa

FIGURE 10.20. Battery-operated temperature and dissolved-oxygen testing instrument that automatically registers readings at desired depths.

It is important to understand that these tests supply only a general analysis of a lake. Such pollutants as pesticides and metals are not identified. A comprehensive lake-monitoring program includes many more tests, taken periodically over a year or more.

In recent years limnology has advanced rapidly as a science because of public concern for water pollution problems. New testing equipment is constantly being developed and marketed. Some is expensive, but it makes the job of monitoring a water-body relatively simple. Less expensive materials are also available. Some are inaccurate and tedious to use. As a result, it is difficult to recommend any one instrument or testing procedure. To save time and money, check with both county and state officials before purchasing test equipment.

TEMPERATURE TESTS

Temperature tests are conducted for several reasons. The most important is to determine the degree of thermal stratification in the lake during the summer. Water temperature has a direct effect on the concentration of oxygen dissolved in the lake. It also directly affects the growth of certain aquatic animals.

As a rule, aquatic plants do better in water warmer than 68° Fahrenheit (20° centigrade), where "rough" (often less desirable) fish can also be abundant. Water cooler than 68° seems to have an opposite effect. The number of plant species is reduced and the population of more desirable fish increases. (See Table 10.1.)

Table 10.1. TEMPERATURE RANGES FOR GROWTH OF CERTAIN ORGANISMS

TEMPERATURE	EXAMPLES OF LIFE
Warmer than 68°F (20°C)	Much plant life; many fish diseases. Fish are mostly bass, crappie, bluegill, carp.
Cooler than 68°F	Some plant life; some fish diseases. Salmon, trout, bluegill, and bass more common.
Cooler than 55°F	Trout; larvae of caddis fly, stonefly, mayfly.

Source: Adapted from U.S. Department of Agriculture, *Teaching Material for Environmental Education,* Stock No. 0101-0234 (Washington, D.C.: 1973).

FIGURE 10.21 Graph Form for Collecting Data at a Water Station and Illustrating Temperature and Oxygen Relationships.

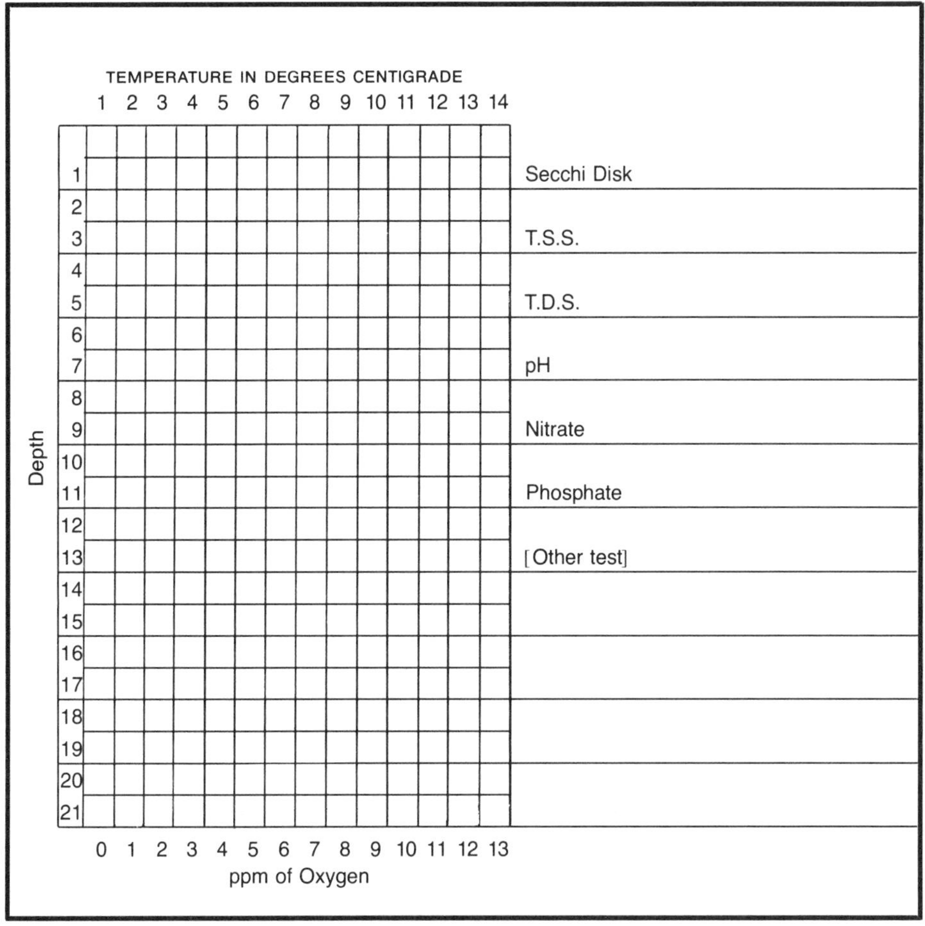

Note: If you wish to convert temperature to Fahrenheit, use the key below:

Centrigrade	1	2	3	4	5	6	7	8	9	10	11	12	13	14
Fahrenheit	33.8	35.6	37.4	39.2	41.0	42.8	44.6	46.4	48.2	50.0	51.8	53.6	55.4	57.2

Temperature readings should be taken at one-meter intervals from top to bottom, especially in the deep sections of the lake. Try to test these areas during the fall, winter, spring, and summer. Temperature readings at each station can be graphed to show the vertical water-column. The graph may illustrate additional tests as well. (See Figure 10.21.)

DISSOLVED OXYGEN

Dissolved oxygen (D.O.) is perhaps the most important indicator of water quality in a lake or stream. Oxygen tests can indicate the effects of oxidizable wastes on water bodies, as well as the rate at which a body can purify itself. Almost all aquatic organisms except for anaerobes depend on dissolved oxygen for survival. (See Table 10.2.)

Table 10.2. DISSOLVED OXYGEN REQUIREMENTS FOR SOME AQUATIC ORGANISMS

ORGANISM	REQUIREMENT
Cold-water organisms (including salmon and trout)	
Spawning	7 ppm* and above
Growth and well-being	6 ppm* and above
Warm-water organisms (including bass and bluegill)	
Growth and well-being	5 ppm* and above

Source: Adapted from U.S. Department of Agriculture, *Teaching Material for Environmental Education,* Stock No. 0101-0234 (Washington, D.C.: 1973).
*ppm = parts per million

If tests suggest that oxygen is consistently low in a water body, all higher forms of life will be affected, and there may be some undesirable side effects. Anaerobic bacteria (bacteria not requiring free oxygen for respiration) can multiply and decompose organic matter, thus liberating as byproducts hydrogen sulfide and methane. This process can be normal in a marsh but undesirable in a lake.

Taking a single sample to determine the amount of dissolved oxygen rarely reflects accurately the overall condition of a water-body. Tests for oxygen should be conducted along with temperature studies because of their relationship. Regardless of the test used for determination of dissolved oxygen, aeration and warming of the water-sample must be avoided, so testing should be done as soon as possible. The scale in Table 10.3 suggests desirable levels of oxygen for lakes. Remember that oxygen can be low at the bottom of deeper lakes during summer when the lake is stratified.

Table 10.3. DESIRABLE OXYGEN LEVELS FOR LAKES

VERY BAD	BAD	FAIR	GOOD	VERY GOOD
Less than 2 ppm*	2–4 ppm	5–7 ppm	8–10 ppm	Over 10 ppm

*ppm = parts per million

TURBIDITY

Turbidity tests measure the amount of suspended and dissolved solids in water. Suspended solids consist of portions of decaying plant and animal matter, silt and clay particles, animal excrement, industrial and domestic waste, and living phytoplankton (minute plants) and zooplankton (minute animals). Dissolved solids consist of phosphates, nitrates, iron, and many other materials.

Suspended and dissolved materials enter the water-body in a number of ways. Runoff from rain or melting snow carries soil particles, litter, and salt from streets, dust from rooftops, and refuse from catch basins and storm drains. Wind and wave action can stir up the bottom of a lake where materials have settled during quieter periods. Varying physical and chemical conditions can give rise to great populations of plankton during different seasons, or cause adverse conditions for them.

Many of these materials eventually settle to the bottom, precipitated by gravitational force. Some are used by plant and animal life as nourishment. The

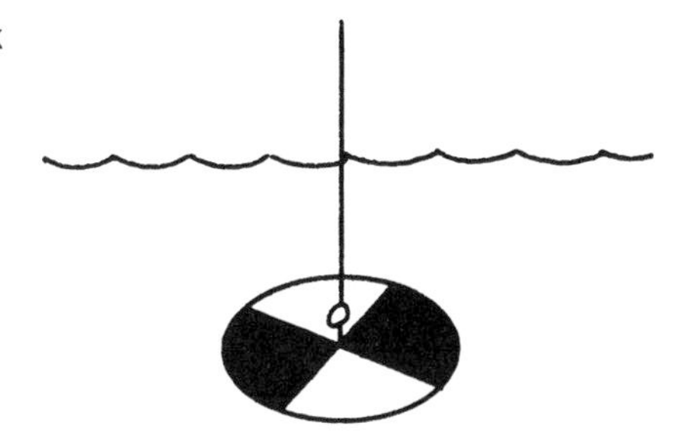

FIGURE 10.22. Secchi Disk

organisms eventually die and parts of their bodies reach the lake bottom. Thus, the degree of turbidity can change from day to day. Aquatic organisms that exist in a water-body are adapted to these normal fluctuations, but rapid increases in turbidity can cause problems. The lake may decrease appreciably in size. Bottom-dwelling organisms can smother. Egg deposits may be covered and destroyed. Gills of fish, crustaceans, and insects can be clogged and rendered nonfunctional. Photosynthesis by plants is reduced, resulting in less dissolved oxygen.

Turbidity is measured in several ways. A Secchi Disk can be used for visibility readings. (See Figure 10.22.) The disk is lowered into the water and the depth at which it disappears from view is then recorded. The disk is pulled up slowly and the depth at which it reappears is recorded. An average is calculated from these two readings. Secchi Disk readings can range from a few centimeters to over 40 meters. Fairly clear Michigan lakes, for instance, have Secchi Disk readings of 3.5 meters (11.5 feet).

Total suspended solids (T.S.S.) can be measured by filtering one hundred milliliters of water through a preweighed filter, letting the filter paper dry, and then weighing the filter again.

Total dissolved solids (T.D.S.) can be determined by using the filtered water from the suspended solid study. Put the water in a preweighed test tube. The water can then be boiled off and the test tube weighed again. The difference in weight will be the dissolved material left behind.

T.S.S. and T.D.S. readings are expressed in parts per million (ppm). Readings can range from less than 25 ppm to 200 ppm and more. Low readings usually indicate a relatively young lake, whereas high readings suggest an older lake. Continued high readings in a young lake may indicate an abnormal influx of pollutants. Because of the variety of substances involved, acceptable levels or ranges are difficult to establish. More important to you as an investigator proposing a development is the identification of the norm upon which future comparisons can be made for the water-body.

pH MEASUREMENTS

A *pH measurement* indicates how acidic or alkaline (basic) a water sample is. The pH value of a lake can be affected by chemical spills, water inflow, weed-control factors, carbon dioxide concentration, and the amount of organic acids present. The pH value usually drops as a lake ages. A lake that is basic when young turns more acidic as organic materials build up. During decomposition, carbon dioxide is released; it readily unites with water to form carbonic acid, a weak acid. This occurs often in lakes with large volumes of organic material in suspension or on the bottom.

Though not a direct result of land change, *acid rain* can be a problem. The story of acid rain begins with industrial and automotive pollutants. The pollutants are carried far from their source and eventually fall to the ground with some form of precipitation. In the process, some of the chemicals unite with water vapor to form weak acids, which, over a period of time, will lower the pH of a body of water. Eventually the pH of a lake will be so low that only those organisms capable of living under extreme acid conditions are able to survive.

A range of 0 to 14 has been established to measure pH values. A pH reading of 0 to 7 units indicates acidic water, 7 to 14 units, alkaline water. A pH value of 7.0 is neutral (neither acidic nor basic). Most aquatic life functions within a narrow pH range. (See Table 10.4.)

Table 10.4 pH RANGES FOR SOME FORMS OF AQUATIC LIFE

	MOST ACID 1 2 3 4 5 6 7 8 9 10 11 12 13 14 MOST ALKALINE
Bacteria	1.0 ---------- 13.0
Plants (algae, rooted)	6.5 ---------- 12.0
Carp, suckers	6.0 ---------- 9.0
Bass, crappie	6.5 ---- 8.5
Snails, clams	7.0 ---- 9.0
Largest variety of animal life	6.5 – 7.5

Source: Adapted from U.S. Department of Agriculture, *Teaching Material for Environmental Education,* Stock No. 0101-0234 (Washington, D.C.: 1973).

NITRATE AND PHOSPHATE TESTS

Nitrogen can be present in water as atmospheric nitrogen, organic ammonia, and inorganic nitrites and nitrates. Of these products, inorganic nitrate is the form most commonly tested for, because it is the form required by plants for syntheses. In water, nitrates come mainly from industrial effluents, sewage, agricultural runoff, animal waste, and decaying plant and animal matter. When these substances enter water, bacteria set about immediately breaking down the more complex organic proteins into inorganic forms of nitrogen. As a result, tests for nitrates can give you a rough idea of the amount of organic material in the water.

This is true for phosphates also. Phosphates are present in three forms: inorganic phosphorus, phosphates tied up in protoplasmic structures, and dissolved organic molecules resulting from the breakdown of protoplasm. Phosphorus enters the water from many of the same sources as nitrogen. Also like nitrogen, the inorganic form is the state most commonly tested for.

Although the total volume of nitrogen and phosphorus can be high in a lake, the inorganic state available for utilization by plants is very small; therefore some limnologists believe that nitrogen and phosphorus combine to set limits on the number of organisms in a lake. Others believe that phosphorus alone is the most critical limiting factor in a water-body,because the ratio of inorganic phosphorus to other elements in the environment is believed to be less than that of the same elements within the living organisms.

Because usable nitrogen and phosphorus are limited, tests should be conducted in the spring and fall, when the temperature is uniform throughout the lake. Because winds will have mixed the water thoroughly, scattered samples collected at this time should give you an accurate picture of the available nitrogen and phosphorus. State your measurements in parts per million. Acceptable levels vary from lake to lake, but present research indicates that levels of phosphorus over 0.015 ppm and 0.30 ppm of nitrogen (nitrate) can contribute to algal bloom. It should be remembered that wide variations do occur. In all cases, your data should serve to establish norms for the water-body being tested for your proposed development.

ANALYSIS OF PLANT AND ANIMAL LIFE

Like all environments, the aquatic environment has representatives of most forms of life. Analysis of the plant and animal species of an area proposed for development aids planners in predicting problems that may result and helps them initiate specific action to avoid them. Fortunately, limnologists have long studied plants and animals of aquatic environments and used them as indicators of aquatic conditions.

Collecting Aquatic Plants and Animals

Collecting aquatic organisms can be both interesting and fun. No matter where you live—even in most deserts—ponds and streams of some type will challenge you with a variety of specimens. Roadside ditches, small temporary pools, and holding ponds are good collecting places.

FIGURE 10.23. Students collecting aquatic plants and animals with a net (seine).

FIGURE 10.24. Vegetative area designed to collect runoff water from the surrounding parking lot. The water that is collected slowly seeps into the soil.

FIGURE 10.25. Homemade equipment for collecting water organisms.

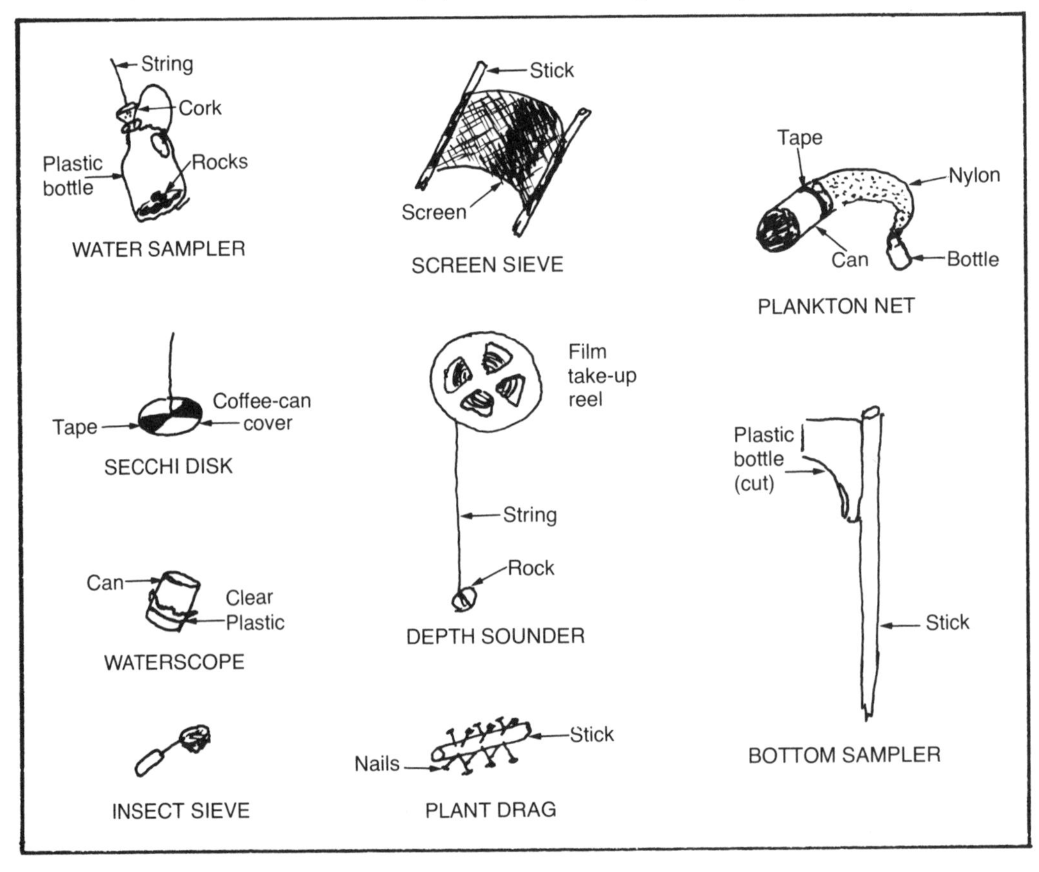

Holding ponds and *sedimentation basins* are small ponds built by developers to slow runoff. These facilities allow the water to soak into the ground or to drain slowly into underground pipes or nearby streams. They also permit sediment to settle out of the water. Because the ponds are often shallow, plants grow rapidly. These areas can provide excellent sites for collecting and observing stages of aquatic succession.

Collecting organisms does not require expensive equipment. Most of the equipment sold by scientific companies can be reproduced at home or in the laboratory. Coffee cans, nylon stockings, window screening, netted football shirts, plastic bottles, and old broom handles can be used to build excellent plankton nets, bottom samplers, and other equipment necessary to collect samples. (See Figure 10.25.)

Identifying and Analyzing Aquatic Organisms

Collecting can be fun, but learning the names of organisms collected and understanding their life styles and the roles they play in the ecosystem are truly challenging tasks. Limnology is a complex discipline requiring years of study. But the animals and plants commonly collected can be identified by their major taxonomic groups, if not their genus and species. Also, many of these common organisms can serve as indicators of water quality or specific aquatic communities.

In the Activities Manual provided for this text, a series of drawings illustrates some of the common plants and animals found in a fresh-water lake or pond. They should be helpful in identifying the specimens you collect. Numerous books and pamphlets are available commercially and from state and federal agencies to help you recognize and learn about aquatic organisms.

CONCLUSION

Procedures for collecting water data described in this chapter can yield baseline data from which to make further studies and comparisons. Over a period of time, this information becomes more valuable because the data will begin to show trends in a water-body's physical, chemical, and biological makeup. From these trends can come judgments about the rate of aging and the degree of pollution that is occurring in that water-body. To manage land wisely, we must understand water in its many conditions. To disregard water in planning environmental changes is to invite trouble. Working *with* water will improve livability wherever it is our privilege to live.

REFERENCES

Amos, William H. *Limnology*. Chestertown, Md.: LaMotte Chemical Products, 1969.

Andrews, William A. *Environmental Pollution*. Englewood Cliffs, N.J.: Prentice-Hall, 1972.

Borton, Thomas E., Marsh, William M., and VanDusen, Peter. *Planning and Management Guidelines for Inland Lake Property Owners*. East Lansing, Mich.: Michigan State University, Institute of Water Research, 1976.

U.S. Department of Agriculture. *Teaching Material for Environmental Education.* Stock No. 0101-0234. Washington, D.C.: 1973.

U.S. Department of the Interior. *Hydrology for Urban Land Planning—A Guidebook on the Hydrologic Effects of Urban Land Use,* by Luna B. Leopold. USGS Circular No. 554. Washington, D.C.: 1968.

U.S. Department of the Interior, Federal Water Pollution Control Administration. *Water Pollution Aspects of Urban Runoff.* Washington, D.C.: 1969.

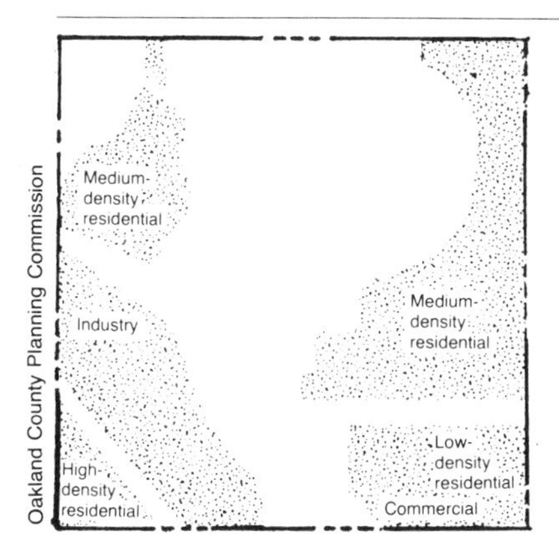

11

METHODOLOGIES FOR PLANNING DEVELOPMENTS

Phase Two of the ecological planning discussed in chapter 3 suggests that planners use a site-planning methodology that will designate areas to be built upon and those that should be modified to some lesser degree or preserved in a natural condition. The data collected in the inventory phase form the basis of the method, but it should also reflect a logical, scientific, and artistically sensitive evaluation of the proposal and establish priorities leading to minimal destruction and maximal benefits. Too often, the placement of buildings, roads, and other developmental features had been guided by rules of convenience and cosmetic, short-term economics. The guiding principle for many housing developers, for example, has been to make the development attractive enough to sell and simple to build. This has led to many of the problems discussed in chapters 1 and 2.

As a beginning planner you must constantly be aware of the dilemma involved in short-term planning. Chapter 2 pointed out that as cities become undesirable places to live, some of the population moves to rural areas; the rural setting thus changes to an urban one, and this often generates long-term social problems. The planner must recognize that people are social organisms, and cities are necessary centers for economic, cultural, and intellectual activities. The planner's problem, then, is to design social and cultural settings that permit the individual and family to maximize their needs without generating future social and environmental problems; the principles of *long-term ecological planning* should prevail in all plans no matter how small the project may be. Your own back yard or school grounds should be no exception!

We have few analytical models or techniques to draw on in our attempt to formulate sound ecological plans. We have, however, made great advances in understanding the various components of the ecological system, such as rock, soil, water, air, plants, and wildlife. To a lesser degree, we appreciate and understand the intricate patterns of relationships that tie the ecological fabric together. More important, we have just begun to understand the necessity of applying this knowledge to our everyday lives. The purpose of this chapter is to consider ways to develop methods of presenting ideas for construction features and site improvements.

COMMUNITY PLANNING TECHNIQUES

Professional planners have historically focused their attention on development strategies for city and county government. This emphasis is the result of state legislation giving local units of government planning power. In an effort to protect the environment, some communities are now writing ordinances that allow greater density than was desired in the past, in the process freeing more land for open space.

A basic concept used by professional planners in the design and layout of developments is the *neighborhood unit concept,* in which planning units are visualized as composed of a central city surrounded by several communities, which in turn are composed of several neighborhoods. The central city is conceived as a complex of people, business, industry, and civic and cultural centers, all connected by a circulation system. The city's function is to provide the basic services required by its citizens and to offer facilities that may not be present at the community and neighborhood level. The one or more communities that would make up the city can supply several neighborhoods with a major shopping facility, a high school or junior high/middle school, and one or more community parks. The neighborhood is the basic planning unit. It contains a population large enough to support one elementary school and a small neighborhood park. Ideally, the street patterns are designed to eliminate through traffic and the school is centrally located within safe walking distance for all school children.

Increasingly more planners have come to realize that for planning to be effective it must be carried out on a regional level but at the same time maintain a focus on the neighborhood. A *region* is a large geographical area that provides a supporting base for one or more cities, towns, or villages. The boundaries of a region may be defined by geographical, ecological, or political boundaries and may contain a regional shopping center, a college, regional parks, and industrial centers. Throughout the United States, regional planning agencies have evolved that focus on a variety of issues. For example, Nashville and Davidson County, Tennessee, have consolidated into a single metropolitan government. Miami-Dade County, Florida, is an example of partial consolidation. (See Figure 11.1 for a neighborhood-regional planning concept.)

In theory, the neighborhood idea is basically sound and has worked to a limited degree, especially with "new communities" that follow the process throughout their development and growth. But for the most part, communities must apply planning principles on a day-to-day basis within the framework of existing development. So, over the years, the basic concept has been modified and expanded to include more amenities for the inhabitants of any existing community or community development. Some of the feelings of alienation in our cities today have arisen because of a loss of a sense of community.

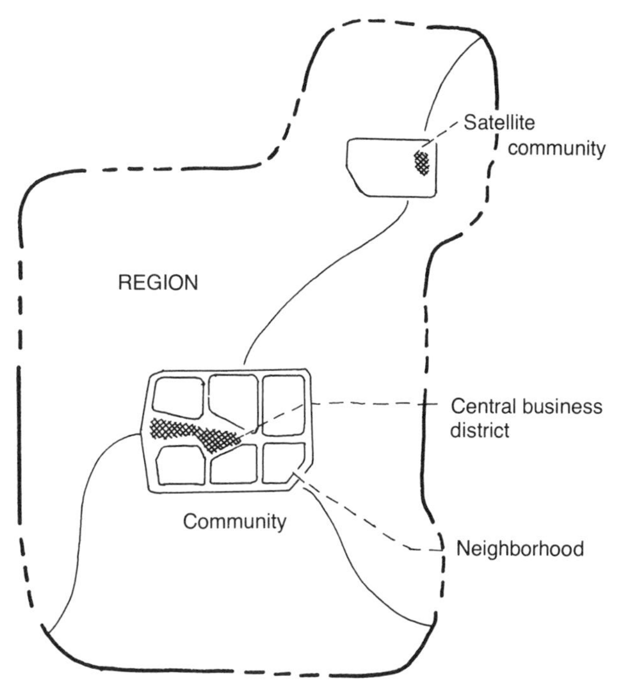

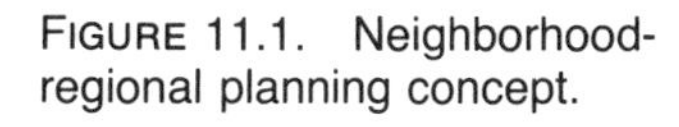
FIGURE 11.1. Neighborhood-regional planning concept.

Increasingly, planners recommend *cluster development.* Most state planning legislation is based on density factors. In the case of residential development, a specified number of people are planned for a given piece of land. Past construction practices too often divide an area into equal parcels. This gave each homeowner an equal share of a development but often increased costs of distributing utilities, destroyed special environments such as wetlands, and created relatively uninteresting, homogenized landscapes. Within the legislated density limitations, many planners are now proposing grouping, or clustering, homes closer together and committing some acreage to common open space. The open space can then be arranged in blocks or patterns for active and quieter recreational pursuits and ecological and aesthetic amenities.

The *planned residential development* (PRD) is a related method of preserving open spaces. Such developments allow a mixture of housing types (single-family, multiple-family, townhouses, and so forth), attached to each other or detached, to be built. (See Figure 11.3). The units are grouped and built two or more stories high, leaving large open areas for common needs. The clustering and mixing preserves the character of abutting land and as much natural terrain and vegetation as possible. Cluster zoning, however, does have disadvantages, especially in relation to making decisions regarding the saved open spaces. How easily, if at all, can a youngster get permission to build a tree house?

Some communities are fortunate enough to have engineering studies outlining the extent of the 2 percent flood plain level for their streams and wetlands. With these data, *flood plain ordinances* can be enacted to control the development of these important resources. Other communities have developed *wetland and waterway ordinances.* These are based on soil studies and vegetation surveys that classify certain parts of the community as restricted to development because of water and soil characteristics.

Though additional regulations are frowned upon by most developers, regulations can produce innovative new ideas. Near Birmingham, Alabama, a 2,700-acre new community, called "Riverchase" was built along the Cahaba River, where development was restricted by a lack of sewers, flood plain regulations, and land that was scarred by dumping and strip-mining activities. Through judicious planning, the scarred land was reclaimed and combined with flood plain to form an eighteen-hole golf course that is partially irrigated with effluent, or waste material, from the development's sewage treatment plant.

Scottsdale, Arizona, a city of 68,000 people, has developed a nonstructural alternative to the traditional concrete drainage system for handling runoff water. A seven-mile-long, 1,227-acre greenbelt dividing the city in two was designed primarily for flood control purposes but also serves as the city's major recreational area, with four major parks, six golf courses, lakes, a trail system, and nature display area.

The principles of clustering and mixing land uses has also been applied to vertical developments. Across the United States and Canada, a number of high-rise projects have been developed that incorporate residential, commercial, and recreational facilities on the same site. In Albany, California, a "vertical village" called Albany Hill was designed on a thirty-six-acre hillside site,

FIGURE 11.2. Cluster development.

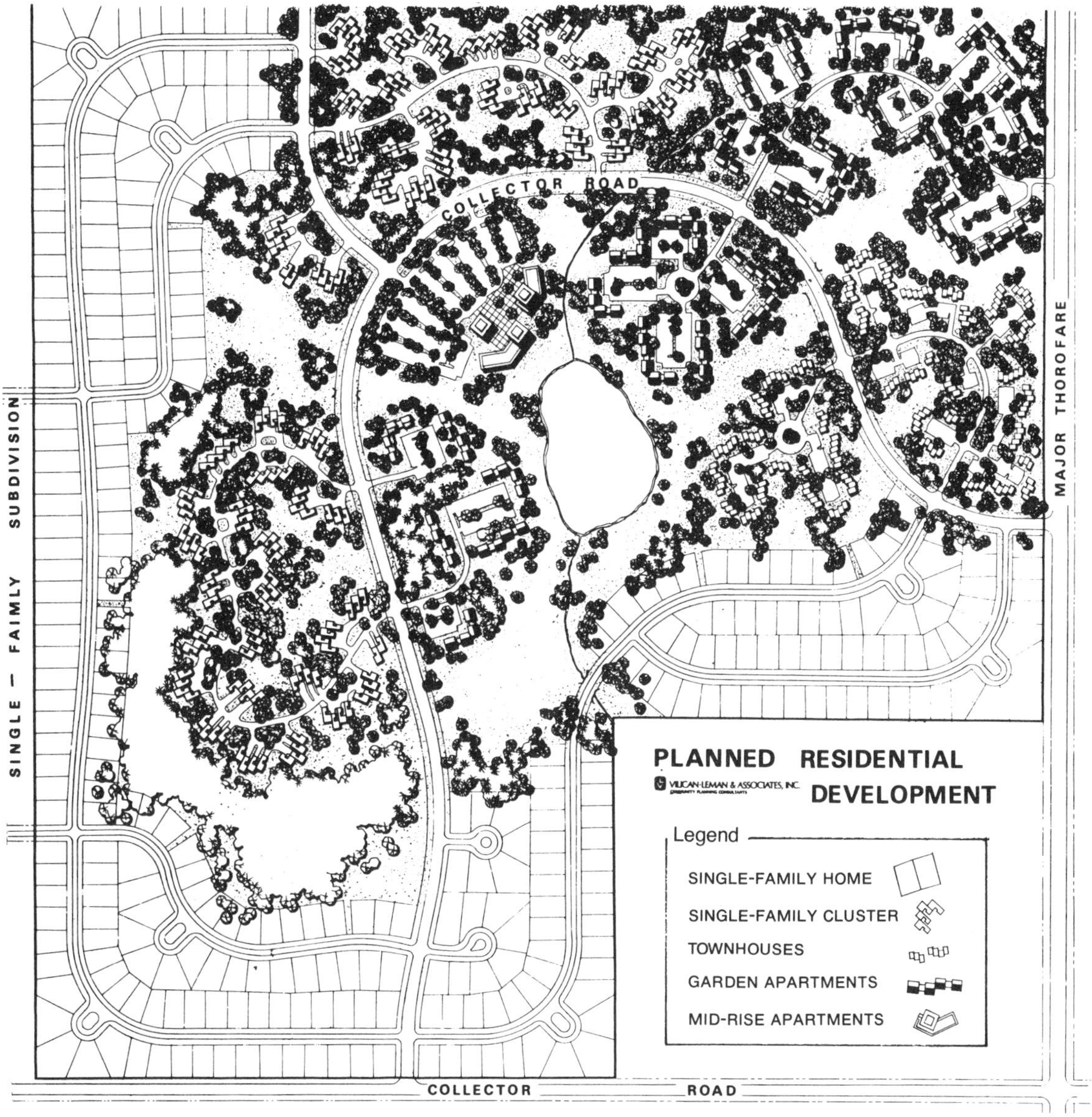

Vilican–Leman & Associates, Inc., planning consultants, Southfield, Michigan

FIGURE 11.3. Layout of mixed single-family homes, single-family clusters, townhouses, garden apartments, and mid-rise apartments in a planned residential development.

with two-thirds of the land devoted to open space for family recreation and nature preservation. To satisfy environmental interests and minimize cost, buildings and enclosed parking facilities were constructed in the flat lower portion of the site.

The futurist planner Paolo Soleri has published plans for massive "vertical cities" that could support a million people per square mile. Soleri visualizes these megastructures as the solution to our ecological problems. Others see only new problems in such plans.

ECOLOGICAL PLANNING TECHNIQUES

One of the earliest planners to recognize the problems of haphazard urban growth was Frederick Law Olmsted. Olmsted is often called the father of American landscape architecture and is best remembered for his role in the creation of Central Park in New York City. His plans for the community of Riverside, near Chicago (1869), and the Boston metropolitan park system (1878) exhibited his strong desire to work in harmony with nature.

More recently, John W. Brainerd has written two excellent guides (*Nature Study for Conservation: A Handbook for Environmental Education* [1971] and *Working with Nature: A Practical Guide* [1973] for helping people analyze the environment before changes are made; these are especially useful for the schoolyard and home.

Possibly the person to have the greatest recent success in using ecological principles in land development is Ian McHarg, author of *Design with Nature* (1969). One of McHarg's analytical techniques is to summarize on overhead transparencies all the social and natural features of value to residents. The inventoried features are given priorities (with colored lines, dots, and so forth) according to their importance to people or to ecological processes. These maps are then laid over one another and projected onto a screen where the lightest areas emerge as the best places to build and the darkest areas suggest preservation. Such features as vegetation, soils, erosion, and drainage values are illustrated on separate maps. (See Figure 11.4.)

In Oregon, on the 1,200-acre Cerro Gordo Ranch, a plan was developed based on an estimation of the amount of runoff water that would be available. The studies, conducted by Professor Charles DeDeuwaerder of Oregon State University, suggested that runoff was the key to good ecological design. Combining soil, slope, vegetative cover, and rainfall data, mathematical studies calculated the amount of water used by plants, the amount that evaporates, and the amount left over for human consumption. These calculations gave planners a rough idea of the number of people the site could maintain.

WILDLIFE-HABITAT-ANALYSIS (WHA) TECHNIQUE

One of the concepts used by planners to provide continuity between various areas in a plan is the concept of linkage. Using such a concept, major elements in a proposal are linked together with natural vegetation or man-made features such as sidewalks and bike paths. An example is the bike paths and walkways that allow children to cycle and walk to schools from surrounding subdivisions. A wildlife-habitat-analysis study uses the linkage concept to establish a continuous system of natural vegetation throughout a development to serve ecological as well as human purposes. It carries out this function by preserving and improving existing natural habitats such as streams, ponds, pockets of vegetation, fencerows, and the edge that these habitats provide. These habitats are then carefully arranged in the design of a development to provide a framework for the roads, buildings, and parking lot required in a development.

Essentially, the wildlife-habitat-analysis technique consists of evaluating the site data collected in the Inventory Phase and applying wildlife management techniques as part of the methodology for planning a development. The tech-

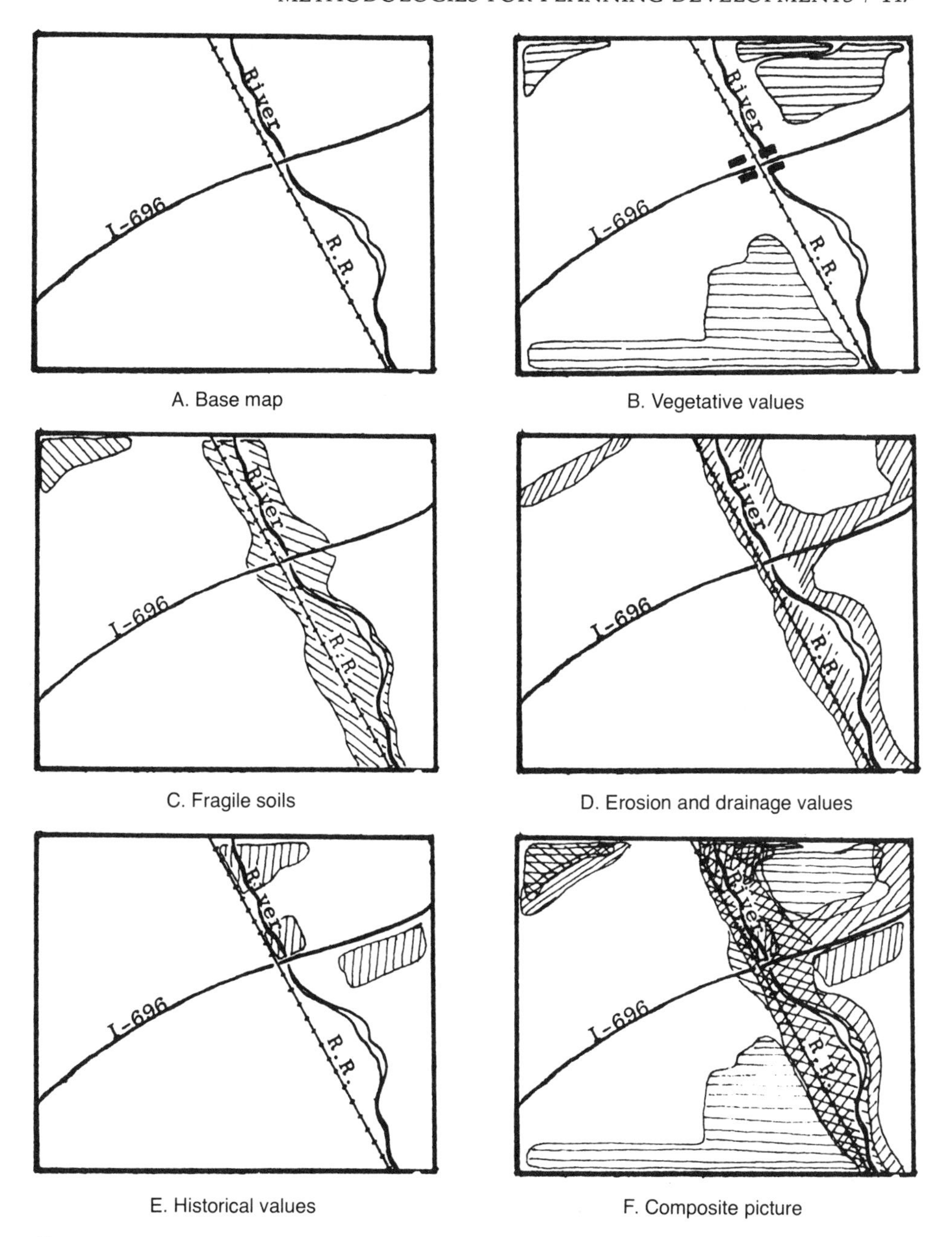

FIGURE 11.4. A simplified example of Ian McHarg's technique of illustrating and analyzing cultural and ecological values.

nique has one advantage over that used by McHarg in that it not only identifies the areas that should be preserved and those that can be built upon with minimum damage to the environment but also establishes a pattern of travel routes around and within a site, thus allowing for additional buffers between different human activities.

To construct a map using wildlife principles, take the following steps.

1. Go over the inventoried data (soil maps, topographical maps, and so forth) and identify all those areas on the site that should be preserved. Outline these on a base map. Try to identify plant-animal communities as *major* and *minor* biotic communities. Major biotic communities are those that provide major human and ecological benefits. Large forested areas, swamps, marshes, and stream beds or flood plains fall into this category. Minor biotic communities consist of such areas as small portions of open fields which are desirable for building but should be maintained to insure adequate vegetative and animal variety. These areas should be scattered as well as small.
2. Study the inventoried data and identify all the topography and vegetation that can function as edge, and identify these areas as *primary and secondary travel lanes*. Primary travel lanes can include old fencerows and forest edges that connect major plant communities. Secondary travel lanes consist of old fencerows, the forest edges, and so forth that connect major biotic communities or simply serve as buffers, screens, or travel routes around the site.
3. Label the rest of the areas as potentially buildable zones, but take other considerations into account. This land will often consist of open fields with good drainage. However, a major drawback to the WHA technique is that it may force development into fields and other agricultural areas. Consequently, take care to identify soil types relative to their importance for agricultural purposes.

In a functional wildlife study carried out on a 377-acre residential/commercial project in Rochester, Michigan, only the primary travel routes were identified, along with the major forest areas and open fields. From the map shown and data from natural resource inventories, it is relatively easy to identify the areas to be preserved and the buildable areas. (See Figure 11.5.)

The Preserve, a 1,100-acre planned residential and commercial neighborhood in suburban Minneapolis, also used linkage-wildlife concepts to achieve a balance between the natural and man-made environments. Existing lakes, marshes, clusters of trees, and a number of old fencerows were incorporated in the neighborhood plan as linkage elements.

GENERAL CLASSIFICATION AREAS

Regardless of the strategy used, the variety of land types on a site should be classified according to their suitability for development. Following are three categories. Other classifications may be developed by the planner and used as the needs arise.

Classification I: Land Unsuitable for Development. Resource areas that should be kept in their existing condition or used for the protection of the public health and welfare. Resources in this classification are wetlands, flood-hazard areas, lakes and streams, seashores, pockets of unique vegetation, and important historic sites.

Classification II: Land Sensitive to Development. Those lands that can be developed with care and forethought. Areas in this classification may include farmland, unique forested areas, certain clay soils, moderately

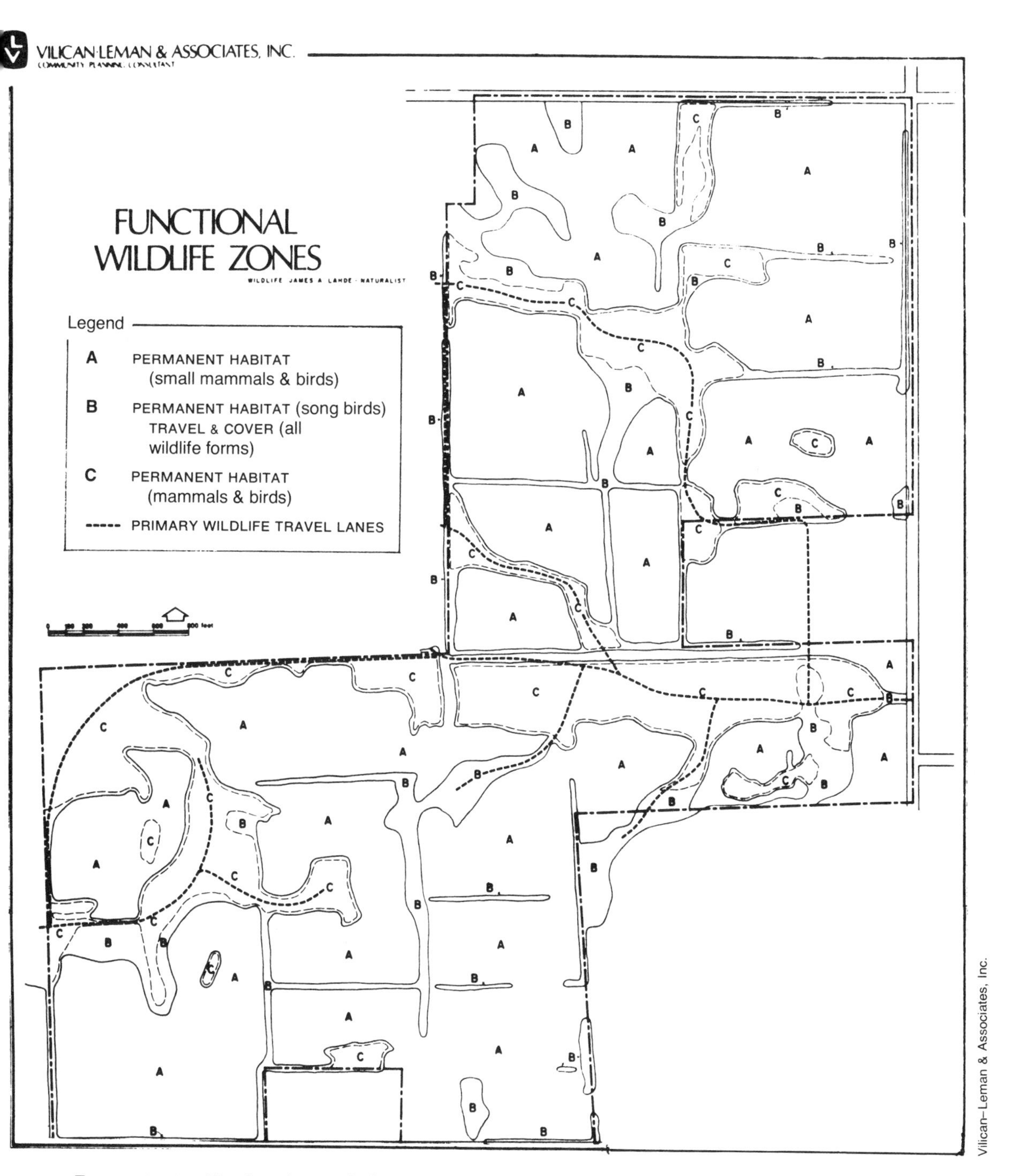

FIGURE 11.5. The location and shape of wildlife zones planned for a variety of housing types on a 377-acre site.

sloping land, and wetlands that because of their small size could be developed.

Classification III: Land Suitable for Development. Should offer little or no resistance to development problems for the inhabitants. Land in this category is usually flat, with good soils.

CONCLUSION

Your method for planning environmental change should define areas that can be built upon with a minimum of disturbance to the resource base and indicate those areas that should be preserved because of their ecological and social values. Once these patterns are established, you can start the job—and fun—of placing the developmental features on the site. The layout of development features, the methods of testing the layout, and the drafting of a final *master plan* for the site are discussed in Chapter 12.

REFERENCES

"Amelia Island Report." Report to the Sea Pines Company, assembled and edited by Jack McCormick and Associates for Wallace, McHarg, Roberts and Todd, July 1971.

Brainerd, John W. *Nature Study for Conservation: A Handbook for Environmental Education.* New York: Macmillan, 1971.

———. *Working with Nature: A Practical Guide.* New York: Oxford University Press, 1973.

Fein, A. *Frederick Law Olmsted and the American Environmental Tradition.* New York: Braziller, 1972.

McHarg, Ian L. *Design with Nature.* Philadelphia: The Falcon Press, 1969.

Soleri, Paolo. *Visionary Cities: The Arcology of Paolo Soleri.* New York: Praeger, 1971.

Vilican-Leman and Associates. "A Planned Community in Avon Township." Southfield, Mich., 1973.

Walker, Dorothy. "Cerro Gordo Experiment." *The Town Forum* (Eugene, Ore.), 1974.

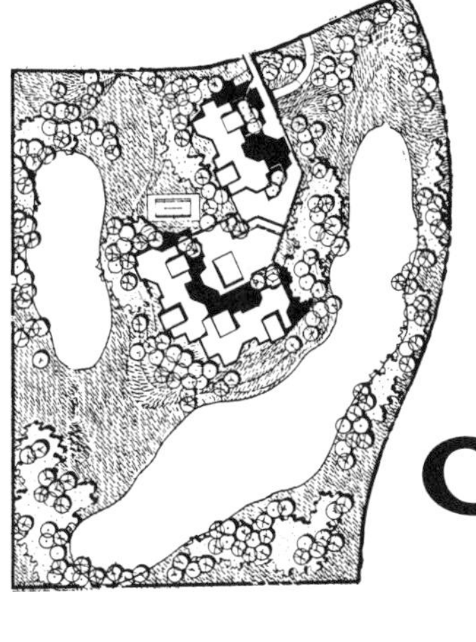

12 CONSTRUCTING YOUR FINAL PLAN

The final step in planning is to construct a *master plan* for your site. This involves: (1) planning the developmental features on the site, or designing the developmental layout; (2) evaluating the layout to insure that environmental, cultural, and developmental needs are met; and (3) making any needed improvements to the site.

In this design phase of your project, you focus on the essential effects you wish to convey. Often these effects can be obtained by your emphasis and articulation of the site's raw materials. Give attention to these basic principles:

1. Seek the site's most suitable features for emphasis.
2. Let the site's resources suggest your layout.
3. Extract the site's full potential.
4. Plan carefully to minimize environmental damages and maximize human benefits.

An example of utilizing the full potential of the site is illustrated in Figure 12.1. A site can have rolling topography with interesting natural vegetation. If the site is leveled, its natural character is lost, but if the natural features are maintained and emphasized in the development, the effects will be rewarding to all involved.

LAYOUT OF DEVELOPMENTAL FEATURES

In this portion of your study, the developmental features (such as roads, ballfields, and buildings) should be laid out on the site within the framework of patterns established in Phase Two. The best way to lay out the developmental features is according to the principles that govern a *conceptual scheme* (theme). For example, it may be decided that a small shopping center should have an "early American" theme. Consequently, this concept might suggest a series of

FIGURE 12.1. A planner should use the full potential of a site. The top drawing illustrates a site with a variety of natural features. In the middle drawing many of the site's features are destroyed, whereas in the bottom drawing development is harmonious with the landscape, resulting in little destruction to the resource base and providing a number of desirable amenities.

small specialty shops situated around a town square with cobblestone streets. (See Figure 12.2.)

The major theme for the layout may have been identified through market studies in Phase One, but it remains for the planner to plot the facilities incorporated in the theme in a creative fashion to establish its character in a functional manner.

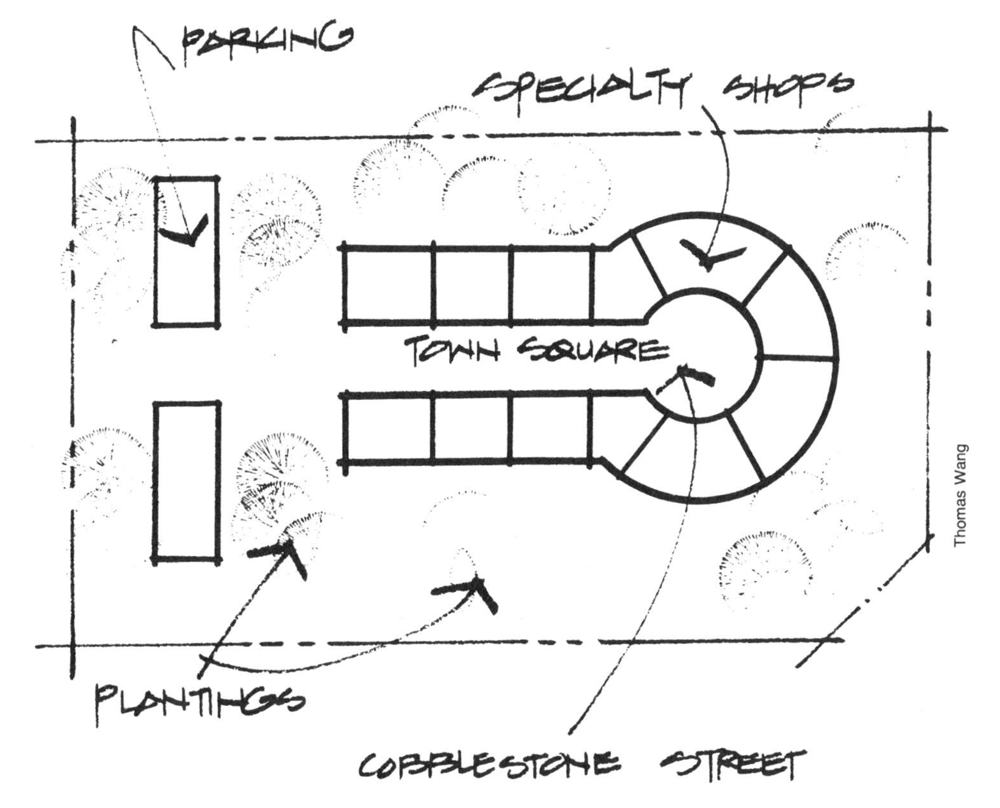

FIGURE 12.2. Layout for developing an "early American" shopping center.

Visual and Spatial Considerations

A *view* is a scene observed from a given point. Many of the landscape features that exist on a site, or those planned for a site, can contribute to or distract from the experience of the viewers. Efforts should be made to identify these visual elements and plan for their enhancement or concealment.

Closely related to the visual element in a design is the spatial element. Much of the art and science of architecture, landscape architecture, and land planning is concerned with *spaces.* Consider the sequence of drawings in Figure 12.3. In the first scene, you might imagine the feelings produced by the landscape to be rather stark, lonely, possibly even frightening. As the space becomes more enclosed the feelings would be more comfortable and secure. Test this enclosure effect by sitting first in the middle of a football field and then under the branches of a large tree. Realize, however, that people may have different feelings about spaces.

Throughout this phase of the study, you as a planner must keep in mind who the primary users of the site will be and the type of structures that will be constructed. If the primary users will be children, then views and scales must reflect their size, feelings, and values. The size, texture, and color of a building also have an impact on the visual and spatial elements. A large brown brick building with straight sides may suggest desert cliffs. A colorful flowering tree

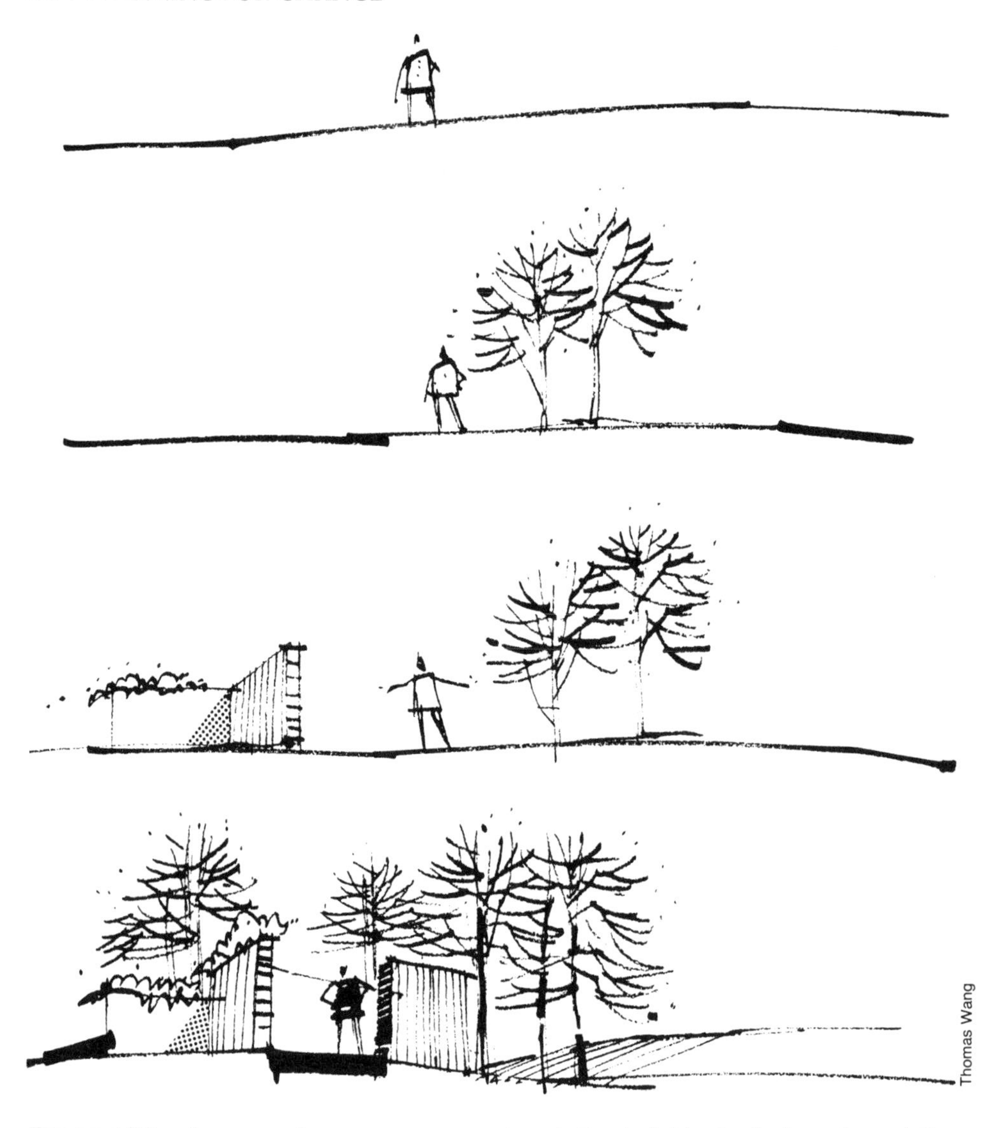

FIGURE 12.3. As space becomes more enclosed, the individual's feelings toward the space may change.

planted beside a drab building can modify the effects of straight lines and single colors.

The effects of wind and sun can have a considerable impact on space also. In Chicago, the onshore breeze from Lake Michigan on the hottest days can lower the temperature by 10° Fahrenheit; so can a sea breeze in Boston. Temperatures in parts of Arizona can vary from 16° Fahrenheit on the roof of a house to 90° at the surface of grass in the shade. Though these examples may be extreme, vegetative and topographic features should be used to control environmental factors wherever needed. Ranchers on the Great Plains know the microclimatological values of their planted windbreaks!

FIGURE 12.4. Small, on-site retention facilities can be attractive and functional. Their design can increase the market value of a development and serve the people who use the site as comfortable spaces to relax.

Runoff Considerations

Many of the problems related to water runoff can be alleviated by your planning. Buildings, parking facilities, and other high-density areas should, whenever possible, be concentrated to retain spaces where the soil's and vegetation's capacity to absorb water can be saved. Below rooftops and on parking lots where water will pour off rapidly, effort should be made to spread the water over large vegetated areas. Where possible, swamps and marshes should be used to collect runoff. Where concentrations of runoff are heavy, natural-looking retention ponds can be designed to hold water. In many older cities and large developments, more elaborate drainage facilities may be required, such as channelization and underground drainage. These techniques should be recommended only as a last resort.

With these basic design principles in mind, you must attempt to position buildings, activity areas, and parking facilities to give the most advantage to the user and reduce the impact on the resource base. (See Figure 12.5.) Circulation

FIGURE 12.5. Try to position buildings, activity areas, and parking facilities to give best advantage to users.

Thomas Wang

patterns must be arranged to move people through your site with minimum effort and maximum benefits. Plants, stones, bricks, and timbers should be used to create interesting spaces, buffer noises, screen unpleasant views, and provide travel lanes for humans and wildlife.

The whole arrangement should be, in the perspective of users, interesting, comfortable, pleasing, and functional. Each development should have its own unique layout. This is especially true of a site plan that is sympathetic to the natural environment. No two natural places are identical.

This portion of the study can be one of the most difficult for you as a young planner, because it demands at least a basic knowledge of the component parts that go into your theme. In addition, it demands creativity and patience to arrange and rearrange the layout so that all the needs are met. A school courtyard planned as a Japanese garden, for example, will suggest that you be familiar with the features that make up such a garden. This will require that you research Japanese gardens and then creatively arrange their elements into a limited space, often with limited funds.

EVALUATING YOUR LAYOUT

Once a conceptual scheme has been laid out, it should be tested. This evaluation process is best accomplished with help from a number of people who have been involved in the study. It is important to have a variety of ideas and viewpoints, because of the large number of variables that must be considered. On big projects today, computers are often used to help establish priorities, but for small projects, the cost of computer time and programing is prohibitive.

On small projects, the patterns established in a wildlife-habitat-analysis study can help alleviate many of the ecological problems involved in the decision-making part of the planning process. But there are many other areas that should be considered.

A checklist can be developed for projects to help you evaluate your layout. The checklist may help bring to mind the many interrelationships that will exist in your proposed development and the conflicts that might result from a proposed change. The checklist represents only broad categories of interest and values. It should stimulate questions about these interrelationships. (See Figure 12.6.)

One simple method of testing a layout is the "let's pretend" method. In this case, an individual must imagine he or she is on the proposed site after it has been constructed, making judgments from critical observation points in the landscape about how effective the design is in achieving the objectives of the theme.

For example, take a look at Figure 12.7, the layout for the back yard of a suburban home. The layout of the active recreation areas includes a pool, tennis court, and horseshoe court. For quiet recreation there is a patio and quiet area with screening and buffer plantings. The views from the back of the home support the theme of the layout, "activity, privacy, and beauty."

Figure 12.8 represents one of the site evaluations carried out for this layout with respect to spatial arrangements of the activity areas and their effects on one another. Circulation patterns were studied along with views from adjacent properties and views between activity areas. These observations were then

FIGURE 12.6

CHECKLIST FOR EVALUATION OF LAYOUT

CATEGORIES OF INTEREST	VALUE JUDGMENTS		
	ADEQUATE	NOT ADEQUATE	COMMENTS
Aesthetic criteria			
Views	____	____	____
Spatial arrangement	____	____	____
Colors	____	____	____
Textures	____	____	____
Ecological criteria			
Noise	____	____	____
Vegetative linkage	____	____	____
Climate	____	____	____
Drainage control	____	____	____
Circulation values			
Vehicular	____	____	____
Pedestrian	____	____	____
Utilities			
Sewers	____	____	____
Water	____	____	____
Electricity	____	____	____
Telephone	____	____	____

FIGURE 12.7. A back yard layout.

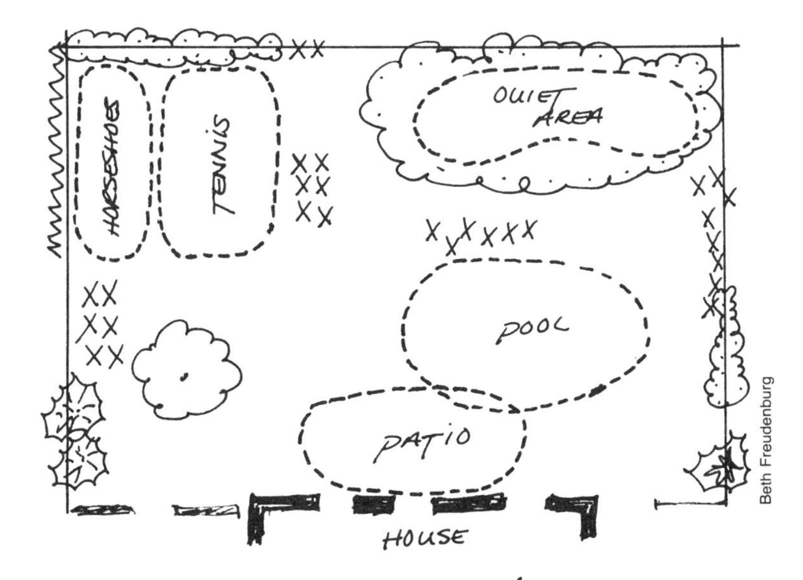

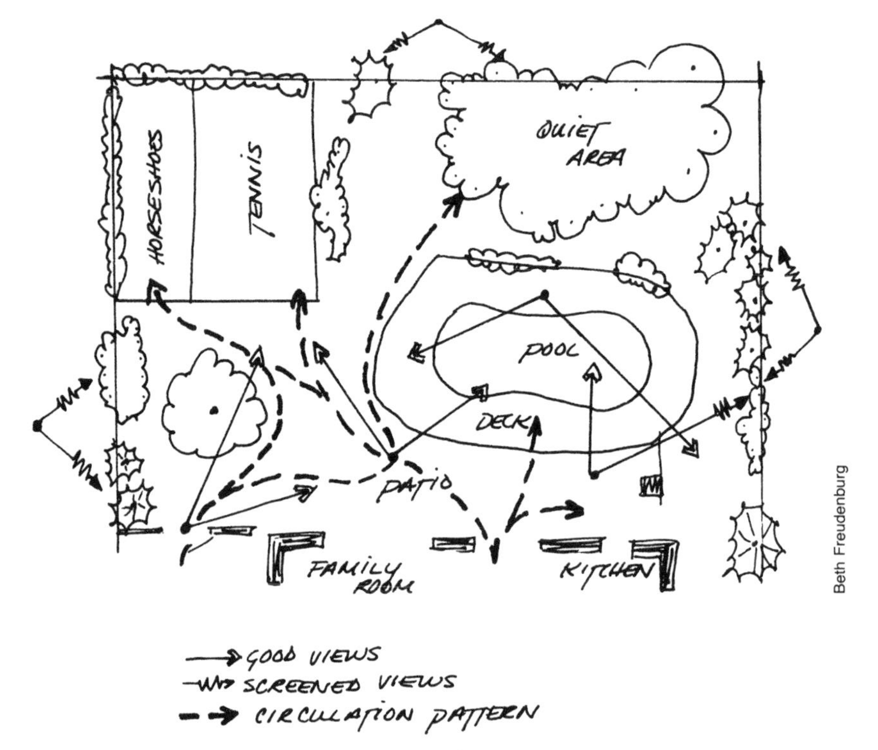

FIGURE 12.8. Checking the spatial relationships, views, and circulation patterns of a layout.

recorded on a map with symbols indicating good and bad effects. Other maps pertaining to other categories of interest could also be constructed—for instance, portraying natural resource values and utilities. When studied together, they could indicate the layout's feasibility.

The fact that value judgments must often be made on two-dimensional plans of three-dimensional construction can often lead to poor evaluations. A more effective but more time-consuming method of testing a layout is to build an inexpensive model of the planned layout. (See Figure 12.9.) The model gives you and others a three-dimensional view of the layout; this can change your perception of how elements in the landscape relate to one another. A model can be made of cardboard, weeds, toothpicks, or any variety of miniature models of appropriate scale sold in model stores. If cardboard "fences" are constructed around the layout with viewing slots cut in them, you can obtain a fairly realistic view of the landscape as it might exist in three dimensions when viewed from normal eye level.

An example of an ecological approach to layout design on a relatively large scale is Chesterfield Village in Saint Louis County, Missouri. The village is an integrated housing, office, and commercial development on a hilly 1,500-acre site. Roads were located along the ridge where they were least disruptive to the environment. The valley slopes were kept forested to preserve the natural character and minimize soil erosion. Culverts, storm sewers, parking lots, and small retention basins were designed to discharge storm water into natural swales, and planted vegetation was used to reduce runoff.

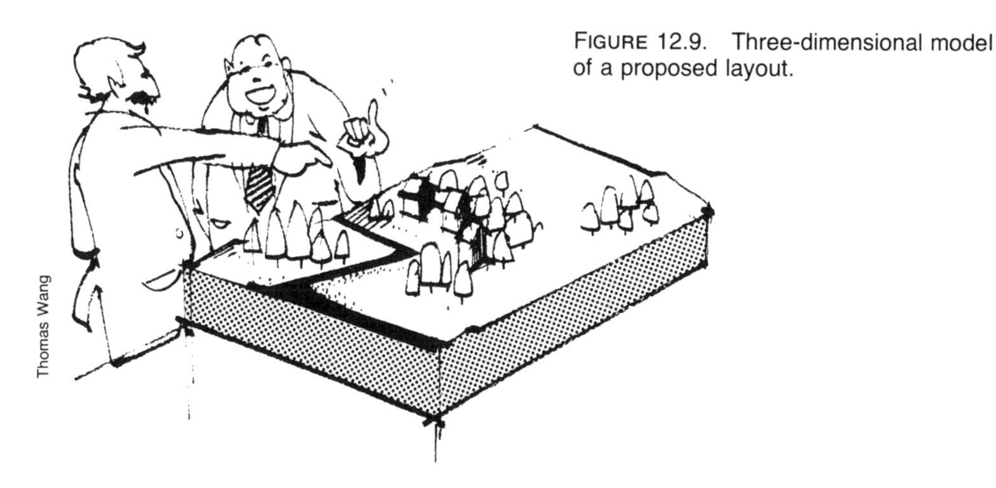

FIGURE 12.9. Three-dimensional model of a proposed layout.

DRAFTING THE FINAL PLAN

The final plan that results from all the preceding studies can be very simple or very detailed, depending on the nature of the project and the people involved in evaluating and constructing the project. There are several drafting methods. The rest of this chapter describes some of the methods available for developing the *master plan.*

Composite Plans

One method of illustrating a final plan is to describe all the features and changes in a plan-view (vertical or bird's eye) drawing. This composite plan should be drawn as realistically and artistically as possible in order to reflect the effort and desires of the planner.

Figure 12.10 illustrates a final site plan for a soda shop. Notice that an attempt has been made to make the buildings, roads, parking facilities, automobiles, and vegetation somewhat realistic. By using any of a variety of media, color can be added to strengthen the impact of the drawing.

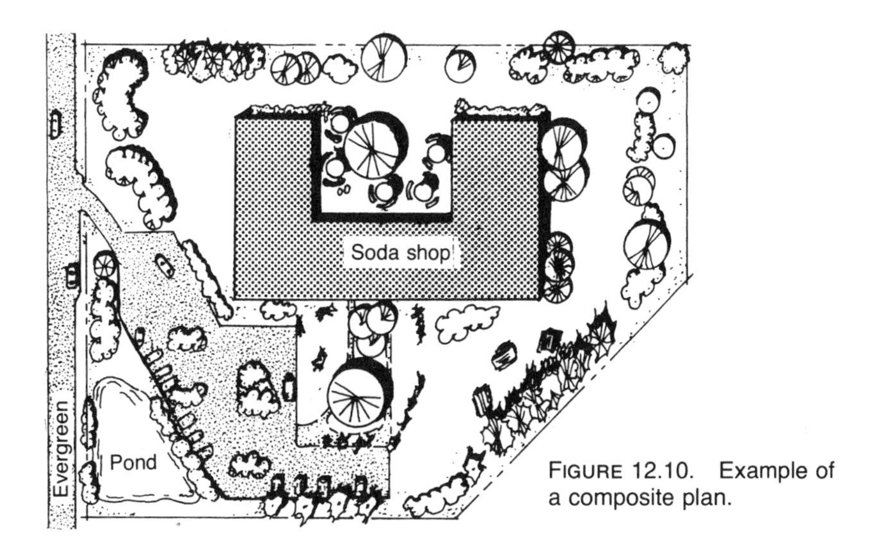

FIGURE 12.10. Example of a composite plan.

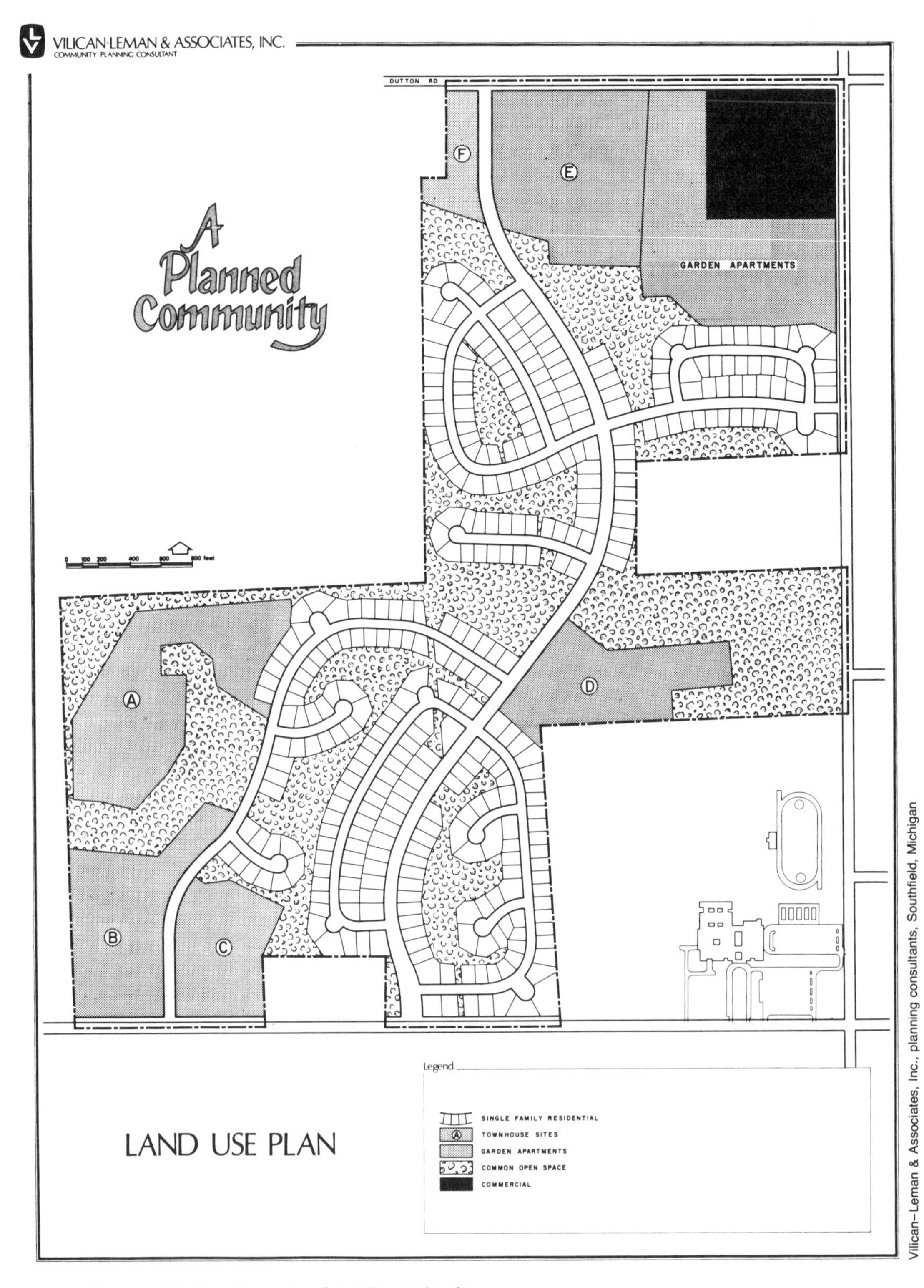

FIGURE 12.11. Example of a schematic plan.

Schematic Plans

Schematic illustrations show the general layout of the development. This type of illustration is often used for large projects where additional studies and drawings will illustrate the project's character. (See Figure 12.11.)

Supportive Illustrations

In almost all cases, it is desirable, even necessary, to include additional drawings with the final plan. Here again, the nature of the drawings is determined by the size of the project and the people involved in the evaluation process.

CONSTRUCTION DRAWINGS

A composite plan or schematic drawing of the total site is too general to show details of features and how they should be constructed. Therefore, construction drawings are important if the plan is to be implemented properly. (See Figure 12.12.) Architectural and engineering drawings usually make up this portion of a study. But even a small plan for a school courtyard or home back yard requires some detailed drawings or sketches, showing, for instance, how to construct a small bridge, pond, or rock garden. Detailed drawings of this nature may re-

FIGURE 12.12. Example of a construction drawing.

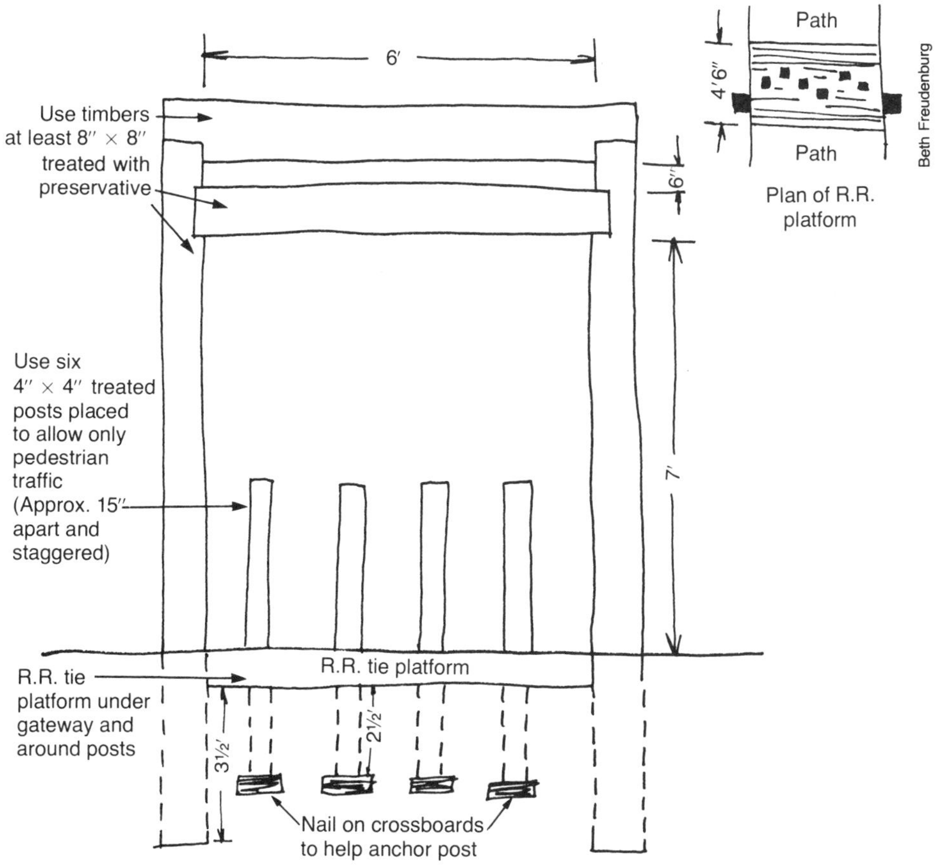

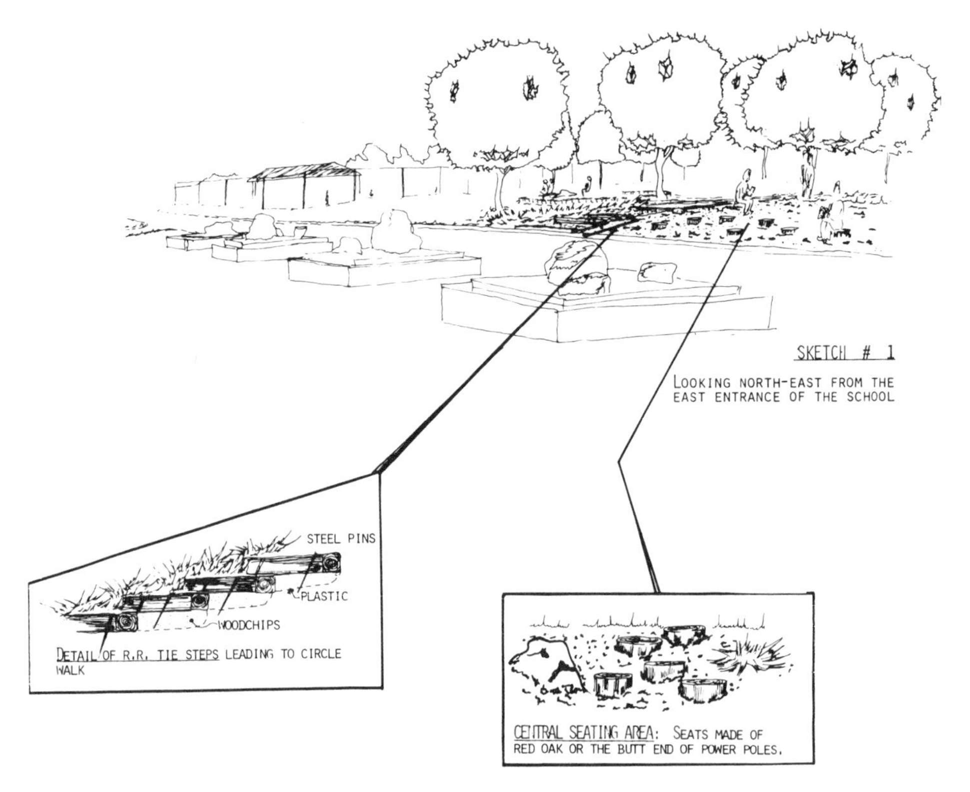

FIGURE 12.13. Example of a perspective view.

quire that you get help from local citizens in the community or local government officials.

PERSPECTIVE DRAWINGS

Sometimes a perspective drawing is included with a final plan. (See Figure 12.13.) A perspective drawing attempts to show what the development will look like in a three-dimensional view from some one vantage point. It is usually colorful and may show human activity throughout the whole project.

CHARACTER SKETCHES

Sometimes character sketches are included with a final plan. (See Figure 12.14.) They attempt to show unique features and human activities that are programed into the site. As a rule, character sketches and perspective drawings are included only when they are useful to "sell" the ideas included in the project to a client or an audience.

MISCELLANEOUS STUDIES

After the final layout has been agreed upon and illustrated, additional studies may be required to improve the site. Every project should consider to some degree environmental-impact studies, landscaping plans, erosion and sedimentation plans, and surface-water-improvement studies. Some states demand ero-

FIGURE 12.14. Examples of character sketches.

sion and sedimentation plans except on some individual home sites. Many developments desire landscaping plans designating the species to be planted.

The National Environmental Policy Act (1970) requires all federal agencies and certain federal projects in conjunction with the states to submit a detailed statement on the impact new projects may have on natural and social environments. The act requires a systematic and interdisciplinary approach in the study. A number of states, from California to Maine, have extended the intent of the national legislation to the state level, so that private developers of large and significant projects are required to submit environmental impact statements for state approval.

Environmental impact studies, landscape plans, engineering studies, and architectural plans require the expertise of professionals in the field and are beyond the scope of this book. Because the ecological planner has made in-depth studies of the natural resources on the site, he or she may be called upon to help make improvements or recommendations on ecologically important features of the plan. This is especially true with vegetational and wildlife modifications.

For example, studies may be carried out that show the number of wildlife zones and major plant communities that will be preserved by the final plan. These studies can help in estimating the degree of environmental impact to the resource base and in pointing out project improvements that should be made. Figure 12.15 shows that most of the larger concentration of trees in the upland, mixed, and lowland hardwoods categories would be preserved if the proposed plan for this development were initiated. It is also easy to observe the location and variety of other vegetational communities and travel routes preserved by the plan. Additional graphs and charts can be constructed pointing out the exact areal percentage of plant communities preserved. (See Table 12.1.)

When the areas that will be affected by development have been located, the planner can make recommendations for their improvement. If the plan is to maintain its ecological orientation, any recommended improvements should be in harmony with the natural character of the plan.

For example, plantings should follow the natural character of a wildlife travel lane. (See Figure 12.16.) Large, sun-loving trees should be planted in the center,

FIGURE 12.15. Environmental impact map.

Table 12.1. PERCENTAGE OF OPEN SPACE AREAS PRESERVED

CLASS	DESCRIPTION	ACRES IN OPEN SPACE	TOTAL ACRES ON SITE	PERCENTAGE SAVED
A	Forbs and grasses	23.7	187.4	13
B	Upland hardwoods	8.9	21.9	41
C	Mixed and lowland hardwoods	46.5	52.0	89
D	Upland shrubs	8.6	92.0	9
E	Lowland shrubs	7.4	21.1	35
F	Marsh-cattail	0.4	1.9	21
	Water	0.3	0.7	43
	New plantings	5.6		
	In disturbed areas*	2.8		
		104.2	377.0	28

*Areas where cut or fill enters into the open space, mainly on roads where grading has to go beyond the road right-of-way.

with shrubs on both sides. A certain portion of the grass and forbs that border the shrubs should be left uncut.

In the case of the 377-acre Michigan site discussed in chapter 11, it was decided that the improvement plantings should consist of trees and shrubs not found on the site. These plants would add a greater variety of color, form, texture, and wildlife food, and cover. (See Table 12.2.)

The travel lanes could be planted with appropriate plants adapted to the soil

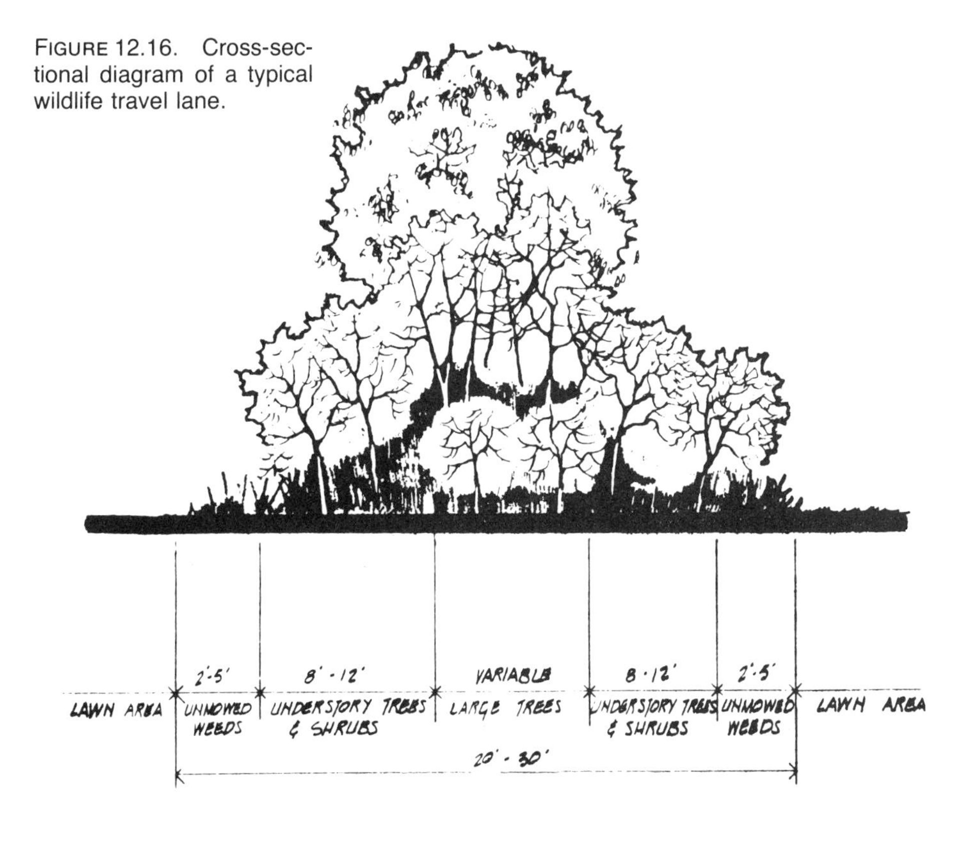

FIGURE 12.16. Cross-sectional diagram of a typical wildlife travel lane.

and water conditions. (See Tables 12.2 and 12.3.) For your region you should do library research and inquire from local professionals what plants they recommend, to supplement the information in this book.

Sometimes requests are made for ideas that may improve wildlife conditions on sites already developed. This is especially true of school sites where many student projects are aimed at wildlife improvement. Following are partial lists of some ideas that may help:

PLANTING PRINCIPLES

1. Avoid too much manicuring of forest and field areas. Weedy fields and strips of sand are necessary for many small mammals and birds.
2. Vary the cover as much as possible; the more varied the cover, the more likely a variety of wildlife will be present.
3. Mix evergreen plantings with hardwoods. Do not plant the same species in large blocks.
4. Favor trees and shrubs with high wildlife value, especially species with heavy seeds, berries, and other fruits.
5. Cluster fruit and evergreen plantings near the forest edge or near fencerows.
6. Maintain fence corners, borders, road median strips, and interchanges with natural vegetation.

IDEAS FOR ENCOURAGING WILDLIFE

1. Provide scattered salt licks, where legal, for larger forms of wildlife.
2. Build bird feeders.
3. Make birdbaths, especially with dripping water.
4. Make brush piles for protection and nesting sites, or as temporary fillers in travel lanes.
5. Save "den trees" (trees with cavities where animals live) in wooded areas to provide homes for woodpeckers, squirrels, and so forth.
6. Erect woodpecker boxes to aid in natural control of bark beetles.
7. Erect squirrel nest boxes in wooded areas.
8. Leave or construct rock piles for smaller forms of wildlife.

IDEAS ON SURFACE-WATER MANAGEMENT

1. Encourage cattail, sedge, and other weedy growth in shallow areas.
2. Provide floating logs or rafts as loafing sites for turtles and birds.
3. Establish waterholes at springs or in seepage areas.
4. Provide potholes and other small openings for water areas for nesting or resting waterfowl.
5. Design and build erect wood-duck nesting boxes.
6. Stabilize stream beds with man-made features or natural plantings.
7. Develop stream improvement devices, such as low rock dams and log dams, deflectors, and flumes.
8. Allow stream-bank vegetation to develop naturally so that it will provide protection for wildlife, shade the stream, and filter runoff.
9. Provide erosion-control devices and plant eroded areas with species good for wildlife food and cover.

Table 12.2 PLANTS FOR ATTRACTING WILDLIFE

TREES	TALL SHRUBS	LOW SHRUBS
Flowering dogwood	Autumn olive	Common winterberry
Crabapple	Russian olive	Almond cherry
American holly	Chinese privet	Japanese barberry
Common laurel cherry	Inkberry	Coralberry
Serviceberry	Silky dogwood	Arrowwood viburnum
Mountain ash	Elderberry	Cotoneaster apiculata
Norway maple	Mulberry	Snowberry
White pine	Possum-haw viburnum	Beautyberry
Norway spruce	Siberian crab	Multiflora rose
Walnut	Hawthorn	Honeysuckle
Austrian pine		Cranberrybush viburnum

Table 12.3. TREES AND SHRUBS RECOMMENDED FOR A SITE IN SOUTHERN MICHIGAN

SPECIES	SUN/SHADE	WET/DRY	FUNCTION	EXAMPLES OF WILDLIFE
Large trees				
Beech	Lt.Shd/Sun	Moist	Food, nesting	Nuts and berries are excellent food for all forms of birds and mammals. Spring foliage is cover for nesting animals and birds. Leaves return a large quantity of minerals to the soil.
Cherry	Sun	Moist/WD	Food, nesting	
Hickory	Lt.Shd/Sun	Moist/WD	Food, nesting	
Red maple	Shd/Sun	Moist/WD	Food, nesting	
Red oak	Lt.Shd/Sun	Moist	Food, nesting	
White oak	Lt.Shd/Sun	Moist	Food, nesting	
Evergreen trees				
Hemlock	Sun	Dry	Food, cover	Squirrels, songbirds (Seeds in the fall and winter are food for squirrels and some songbirds; also good year-round nesting cover for wildlife.)
Red cedar	Sun	Moist/Dry	Food, cover	
White pine	Sun	Dry	Food, cover	
White pine	Sun	Dry	Food, cover	
Small trees				
Autumn olive	Sun	WD	Food	Songbirds
Crabapple	Sun/Shd	Moist/WD	Food	Mammals, game and songbirds
Flowering dogwood	Sun	WD	Food	Songbirds
Hawthorn	Sun	Dry	Food, cover	Mammals, game and songbirds
Mulberry	Sun	Moist/WD	Food, cover	Mammals, game and songbirds
Rocky Mountain ash	Sun	WD	Food, berries	Songbirds
Serviceberry	Sun/Shd	Mod.Dry	Food, berries	Song- and game birds
Winterberry	Lt.Shd	Wet/Moist	Food, cover	Songbirds
Witch hazel	Sun/Shd	Moist/WD	Food, cover	Mammals, game and songbirds
Shrubs				
Buffalo berry	Sun	Moist/WD	Food, cover	Mammals, game and songbirds
Cranberry bush viburnum	Sun/Shd	Moist	Winter food	Game and songbirds
Elder	Sun/Shd	Moist/WD	Food	Mammals, game and songbirds
Gray dogwood	Sun/Shd	Moist/WD	Food, cover	Mammals, game and songbirds
Multiflora rose	Sun	WD	Food, cover	Mammals, game and songbirds
Ninebark	Sun/Shd	Moist/WD	Food, cover	Game and songbirds
Red osier dogwood	Sun/Shd	Moist	Food, cover	Mammals, game and songbirds
Silky dogwood	Sun/Shd	Mod.Dry	Food, cover	Mammals, game and songbirds
Wild rose	Sun/Shd	Moist/WD	Food	Mammals, game and songbirds
Vines				
Bittersweet	Sun/Shd	WD	Food, cover	Mammals, game and songbirds
Grape	Sun	WD	Food	Mammals, game and songbirds
Honeysuckle	Sun/Shd	Dry/WD	Food, cover	Game and songbirds
Virginia creeper	Sun/Shd	Moist/WD	Food	Game and songbirds

Key to abbreviations: Lt.Shd = Light Shade; Shd = Shade; WD = Well Drained.

John Brainerd

FIGURE 12.17. Rock dams made by students to slow the flow of water.

Virginia Finney

FIGURE 12.18. Your goal should be to create spaces that will be functional and enjoyable for decades.

CONCLUSION

The term *animated* is used by planners to describe a plan that does not just cover the physical and economic aspects but also involves the emotional aspects—the way people perceive and enjoy, about a development. It is a good term and a necessary concept but used in this context the emphasis is still on *people.*

The planner should attempt to arrive at a layout concept for the future that looks past the fads of the present and creates spaces that will be functional and enjoyable for decades. A plan of this nature must have a strong ecological base, *animated* to all life and respectful of all men, for only then will there exist the relative harmony necessary for continued existence on earth.

REFERENCES

Laurie, Michael. *An Introduction to Landscape Architecture.* New York: American Elsevier, 1975.

Lynch, Kevin. *Site Planning.* Cambridge, Mass.: M.I.T. Press, 1971.

Martin, Alexander C., Zim, Herbert S., and Nelson, Arnold L. *American Wildlife and Plants: A Guide to Wildlife Food Habits.* New York: McGraw-Hill, 1951.

Michigan Department of Agriculture. *Michigan Soil Erosion and Sedimentation Control Guidebook.* Lansing, Mich.: 1973.

Simonds, John O. *Landscape Architecture.* New York: McGraw-Hill, 1961.

U.S. Department of the Interior. *A Procedure for Evaluating Environmental Impact,* by L. P. Leopold, F. E. Clarke, B. B. Handshaw, and J. R. Balsley. USGS Circular No. 645. Washington, D.C.: 1971.

U.S. Environmental Protection Agency. *Environmental Impact Statement Guidelines.* Washington, D.C.: U.S. Government Printing Office, 1972.

Mary Douse

13

PLANNING FOR CHANGE: A NEW ORDER

A town is saved, not more by the righteous men
in it than by the woods and swamps that surround it.
A township where one primitive forest waves
above, while another primitive forest rots below,—
such a town is fitted to raise not only corn and
potatoes, but poets and philosophers for the coming ages.

—Thoreau
Excursions, 1862

Our country has passed through a pioneering stage, but the pioneering philosophy of freedom to do as one wants with the land and its resources is still strong. In seeking better ways, our planning strategy has been to use zoning laws and building regulations. This approach has no doubt been necessary, but we have now reached a point where a more positive approach is necessary. We must start to examine new strategies more in keeping with scientific discoveries, exploding populations on a planet with limited resources, and cultural demands.

The world is changing rapidly. Because of population increases and technological advances, urban centers throughout the world are experiencing rapid growth and just as rapid decay.

Many of our social disorders and ecological problems result from shifting economies and changing life styles. Population figures for 1970 and 1980 show that large metropolitan American cities, except for those in the Southwest, lost a large portion of their population. Some of this population migrated to the Southwest where job opportunities are more plentiful. Many planners suggest that lenient zoning laws and environmental regulations are partially responsible, along with the Sunbelt's climatological advantages, for this shift of industry and people. They also suggest that uncontrolled development and a limited water supply will result in predictable urban problems there as elsewhere.

Many people are leaving the city for noneconomic reasons. They are willing to accept lower incomes in a rural setting in order to live in a rural environment.

Surveys indicate that reasons for moving include dissatisfaction with city living, with crime, drugs, poor education, taxes, and racial problems. Whether or not the reasons are valid, the trend represents a significant loss to the cities.

Very often these noneconomic migrants relocate near small towns in the same state. The initial impact is minimal because the immigrants often settle near lakes, streams, and other picturesque places. The first complaints come from the farmers who feel the effect of inflated land prices through increased property taxes. Additional conflicts arise as local governments are pressured to improve services.

On a global scale, given the trends in population growth and the existing political atmosphere, the prospects for achieving stable urban growth patterns in the near future are questionable. Some pessimists predict catastrophe. Technological optimists predict magnificent large cities scattered across the landscape, capable of satisfying all of people's needs and operating in harmony with nature. But the short-term reality suggests the existing pattern of urbanization, of decay of older cities and shifting populations, will continue.

The future seems dim only when we measure it along existing lines of development. This textbook is one of hope. The reason for hope lies in the fact that change will occur no matter what forces prevail; and with the possibility of planned, rational change, people may continue to enjoy their place on earth. As Erick Fromm pointed out, "Hope is neither passive waiting nor unrealistic forcing of the future."

There is time for change. But we *must* plan for change. We must consciously and deliberately continue our evolution toward a new system of physical order. All the ingredients are present. Scientists have supplied us with the principles of ecology to serve as a framework for change. Computers give us the capability

FIGURE 13.1. The world is changing more and more rapidly.

Barbara Brainerd

Barbara Brainerd

FIGURE 13.2. Nature can supply us with many models for changing our environment. Small beaver dams create habitat for a number of animals that would not live along the stream; they act also as retention ponds for water runoff.

to store, retrieve, and analyze vast knowledge. And communications satellites permit instant distribution of this knowledge to all corners of the earth.

The only ingredient lacking is a systematic approach to the problems. Every natural environment supplies us with models to follow: the meandering river; the spongy forest soil protected by a canopy of vegetation; the beaver ponds scattered along a small stream. The meandering stream should be left to meander. Small roof top collectors, storage tanks, and attractive retention ponds can replace the beaver ponds.

We must begin at the ground level, with the individuals who make the decisions. Few environmental problems are national or global. The total effect may be reflected on a national scale, but the problems are generated and administered at the local level.

The purpose of this book is not to make environmental planners or professional planners of anyone. The purpose is to increase awareness of the existing planning processes and their problems and to provide an understanding of reasonable ecological alternatives. It shows that the world is a complex system and that the knowledge we strive for can be applied to achieve a better environment for all.

As more Americans become aware of land-use problems, their knowledge of ecological planning may permit their participation in the solutions to these problems. The planning process is, in reality, life. Change cannot be stopped. Whether we like it or not, our community will be different tomorrow from what it is today. And in a democracy, if you and I do not work toward a high quality of environment, we simply will not get it.

In the planning process there is room for everyone. *You* can help guide change in your community. Because the field of environmental quality is so large, some of us will be directly involved through our chosen professions. All of us will be involved simply through living. There are many ways to help. We should all be familiar with local and state planning and zoning laws. If your community has a master plan, learn what it is and how it is being implemented. If your community lacks a plan, work for the adoption of a comprehensive planning program.

Encourage your city or rural engineers and planners to adopt programs based on sound ecological principles. For example, community standards and regulations for water runoff should be such that the natural character of the community will be maintained. Wherever possible, try to eliminate concrete pipes and channelization as a means of handling water. Also, encourage local, county, and state officials to adopt ecologically sound plans and policies that help control density and protect farmlands, wetlands, and waterways.

Most important, practice comprehensive planning as a way of solving land-use issues. Schools can be an excellent place to practice. The whole school site or a portion of it can be used as the subject for drawing up plans. The process of planning can be carried out by one class or by many classes, including drafting classes, industrial arts classes, math classes, science classes, and social studies classes.

Biology, physics, and chemistry classes can conduct air, soil, and water studies. Social studies classes can help determine the impact of a development

FIGURE 13.3. There are many fine examples of good planning in every community.

on neighborhood residents. For example, a plan may call for lights to be erected on a baseball diamond. A questionnaire can be designed to find out the impact these lights will have on abutting property owners, and subsequent recommendations can be made to minimize any harmful effects.

Next, carry out an environmental or ecological planning project on a development that is proposed for the community. Invite the city engineer, city planner, building officials, and developers into class to discuss the project. If possible, get permission to conduct studies on the site. Analyze your findings and make recommendations to the planning commission at the scheduled public hearings.

A few final words of caution. Do not be in a hurry to see your city grow or a plan take form. In the rush for tax dollars, many long-term social benefits can be lost. And never subscribe to the notion that a marsh must be filled in because it is "only a breeding ground for snakes and mosquitos." The mosquito, snake, and marsh all play a role in the complex ecological system of which we are only a part. We may not understand or appreciate their value, but ignorance is no excuse for their destruction. Aldo Leopold, who is considered the father of modern wildlife management, once wrote:

> The last word in ignorance is the man who says of an animal or plant: "What good is it?" If the land mechanism as a whole is good, then every part is good, whether we understand it or not. If the biota, in the course of aeons, has built something we like but do not understand, then who but a fool would discard seemingly useless parts? To keep every cog and wheel is the first precaution of intelligent tinkering.

REFERENCES

Borton, Thomas E., and Warner, Katharine P. "Involving Citizens in Water Resources Planning: The Communication Participation Experiment in the Susquehanna River Basin." *Environment and Behavior,* 1975.

Craik, Kenneth H. *New Directions in Psychology.* Environmental Psychology, vol. 4. New York: Holt, Rinehart and Winston, 1971.

Delbecqu, Andre L., Van de Ven, Andrew, and Gustafson, David H. *Group Techniques for Program Planning.* Glenview, Ill.: Scott Foresman, 1975.

Gussow, J. D. *The Feeding Web: Issues in Nutritional Ecology.* Palo Alto, Calif.: Bull, 1978.

Leopold, Aldo S. *A Sand County Almanac.* New York: Oxford University Press, 1966.

Turk, Amos, Turk, J., Wittes, J. T., and Wittes, R. *Environmental Science.* Philadelphia: Saunders, 1974.

U.S. Department of Agriculture, U.S. Forest Service. *A Guide to Public Involvement in Decision Making.* Washington, D.C.: U.S. Government Printing Office, 1971.

Wagner, T. P., and Ortolano, L. "Analysis of New Techniques for Public Involvement in Water Planning." *Water Resource Bulletin,* 1975.

Webb, Kenneth. *Obtaining Citizen Feedback: The Application of Citizen Surveys to Local Government.* Washington, D.C.: Urban Land Institute, 1973.

INDEX